Kinetics of Ziegler-Natta Polymerization

Kinetics of Ziegler-Natta Polymerization

TOMINAGA KEII

Professor, Department of Chemical Engineering
Tokyo Institute of Technology
Tokyo, Japan

KODANSHA
TOKYO
CHAPMAN & HALL
LONDON

KODANSHA SCIENTIFIC BOOKS

Published by

KODANSHA LTD.
12-21 Otowa 2-chome, Bunkyo-ku, Tokyo 112, Japan
and

Chapman & Hall Limited
11 New Fetter Lane, London EC4P 4EE, England

PRINTED IN JAPAN

STANDARD BOOK NUMBER: 412 11730 4
KODANSHA EDP NUMBER: 3043-247160-2253 (0)

Distributed in the U.S.A. by

HALSTED PRESS, A DIVISION OF
JOHN WILEY & SONS, INC., NEW YORK

To my teacher, Dr. Juro Horiuti, and my parents and wife,
Yosaku, Matsuko and Kuniko Keii

Preface

In 1963 Professors K. Ziegler and G. Natta were awarded the Nobel prize for their discovery of a new type of catalysis, which has become known as Ziegler-Natta catalysis. Before this, three other scholars in the field of catalysis had been awarded the same prize; W. Ostwald (1909) for his work on catalytic action, P. Sabatier (1912) for work on the hydrogenation of organic compounds by metal powders, and F. Haber (1918) for his ammonia synthesis. The results of these men had a profound effect on the science of catalysis, and in the same way, the discovery of Ziegler-Natta catalysts has had an outstanding influence on the development of basic and applied catalysis. This discovery has revealed a new type of catalytic action; stereospecific catalysis, new types of polymers; the stereoregular polymers, and a new type of catalytic material; the organometallic complex catalysts. The resulting development of high density (linear) polyethylene and isotactic polypropylene has had a major impact on the world plastics industry.

The success of Ziegler-Natta catalysis has induced important changes in the ideas and philosophy of many chemists and chemical engineers. The industrial application of these specific but highly unstable catalysts overthrew the traditional idea that technical or industrial catalysts must be mechanically strong and have long lifetimes. Research chemists now believe that a wide range of novel catalyst systems may be derived from organic materials which were previously considered too unstable to have any practical application.

There have been a number of criticisms concerning the progress of research on these new catalyst systems. For instance, "Despite the large amount of work carried out, scientific knowledge has lagged behind the huge amount of technological developments in the field of Ziegler-Natta polymerizations" (L. Reich and A. Schindler, Polymerization by Organometallic Compounds), and "The mechanism of this reaction was a subject of extensive investigation and speculation and constant revision" (H. Pines, Advances in Catalysis, vol. 19). However, progress in research into Ziegler-Natta catalysis has in fact been rather rapid, as is apparent from a comparison with the history of research on olefin hydrogenation or ammonia synthesis. It seems to the present author that an important characteristic of current research is an emphasis on the pure organic chemistry of Ziegler-Natta systems. However, in complex heterogeneous systems such as these, kinetic data can yield a considerable volume of information. A knowledge of the rate laws, activation energies and related data are essential in formulating models or mechanisms for these reactions, and are also necessary from the point of view of industrial process design.

The object of this book is to present a comprehensive discussion of the experimental data on the basis of heterogeneous kinetics. Related aspects of Ziegler-Natta catalyst systems such as alkylation, hydrogen exchange, ESR spectra of the system, effects of grinding the catalyst and monomer diffusion effects are also taken into consideration. The models which have been proposed for the reaction mechanism are reviewed critically. It is hoped that this book will be useful to professional chemists, chemical engineers and graduate students in chemistry and chemical engineering, as a typical modern example of the kinetic treatment of complex reactions.

The author gratefully acknowledges his indebtedness to Drs. I. Okura and Y. Doi, of the author's laboratory, for their help in searching the literature, and also to Mr. W.R.S. Steele of Kodansha Scientific for emendation of the English text. The book is dedicated in addition to the following doctors and masters who worked on Ziegler-Natta polymerization in the author's laboratory, as graduate students of Tokyo Institute of Technology. M. Taira (Osaka Gas Co.), K. Soga (Assistant Prof., Kyoto University), M. Aoyama and H. Koga (Toray Industries Co.), Y. Otsuka, T. Akiyama and S. Go (Mitsubishi Chem. Co.), T. Kato (Mitsui Petrochem. Co.), N. Saiki and T. Suzuki (Teijin Co.), A. Takahashi (Japan Carbide Co.), T. Ishida (Chiyoda Chem. Eng. Co.), A. Kojima and Y. Yoshimoto (Dainippon Ink Co.), T. Matsuda (Assistant Prof., Saitama University) and Y. Ono and S. Kanetaka (Tokyo Institute of Technology).

Tokyo
January, 1972

TOMINAGA KEII
Tokyo Institute of Technology

Contents

Chapter 1

Introduction

Ziegler-Natta catalysts have been defined as the products of reaction between two different groups of metal compounds; (1) compounds of transition elements of groups IV to VIII, known as the catalysts, and (2) compounds such as the hydrides, alkyls or aryls of elements of groups I to IV, termed the cocatalysts.[1)] The important characteristics of Ziegler-Natta catalysts lie in their ability to produce stereoregular polymers, i.e. non-branching polymers from ethylene, and isotactic polymers from α-olefins such as propylene. This stereospecificity has given Ziegler-Natta catalysts considerable scientific and industrial importance.

Because of the complex behavior of Ziegler-Natta catalysts, almost all studies, with the exception of the pioneer work of Natta and Pasquon,[2)] have been directed towards the pure organic chemistry of these compounds with little consideration of kinetic effects. Although this approach can be effective in the case of homogeneous catalysts, kinetic studies become essential for heterogeneous catalysts such as many Ziegler-Natta systems. From kinetic data we can determine the rate law, activation energy, number of polymerization centres, average lifetime of the growing polymer chain and other data, all of which are important in the formulation of models or mechanisms for the reactions. Further refinement of these models may then be made by other techniques such as spectroscopy and organochemical studies. From a practical point of view, the kinetic data provide important basic information for process engineering (process design, reactor design, etc.)

1.1 ZIEGLER-NATTA CATALYSTS AS HETEROGENEOUS SYSTEMS

The typical Ziegler-Natta catalysts on which most studies have been concentrated are combinations of triethylaluminium ($AlEt_3$) or diethylaluminium chloride ($AlEt_2Cl$) with titanium tetrachloride ($TiCl_4$) or trichloride ($TiCl_3$), though combinations of the chlorides of many transition metals, e.g. V, Cr, Zr, with organometallic compounds based on Li, Be, Mg, Tl and In have also found some application. These catalysts are usually heterogeneous systems, having a solid phase coexisting with a liquid phase. Further support for their classification as heterogeneous catalysts is provided by the observation that polymerization occurs mainly on the surface of the solid phase; i.e. the reaction loci are situated on the solid surface.

However, the behavior of Ziegler-Natta catalysts is more complex than that of inorganic heterogeneous catalysts such as metals or metal oxides. Thus, as pointed out by Ambroz *et al.*,[3] a remarkable amount of chlorine is removed from $TiCl_3$ during propylene polymerization using a $TiCl_3$-$AlEt_3$ catalyst. Catalyst changes of this type have not been observed in the field of inorganic heterogeneous catalysis. Historically, Ziegler catalysts based on $TiCl_4$ were the first to be investigated, and for some time these were treated as homogeneous catalysts, so for both the above reasons many workers were initially reluctant to treat the Ziegler-Natta catalysts as heterogeneous systems. The recognition of their heterogeneous nature has now encouraged the development of kinetic models to elucidate the reaction mechanisms, along traditional lines of approach.

1.2 REACTION SYSTEMS AND INFLUENCING FACTORS

The polymerization systems usually used in laboratories may be classified into two types; closed and open systems. Polymerization in a dilatometer or a closed autoclave occurs under constant volume, and these are referred to as closed systems, whereas devices for polymerization under constant pressure are known as open systems. The use of a dilatometer by Cooper *et al.*[4] (Fig. 1.1) and an autoclave by Kagiya *et al.*[5] provide typical examples of polymerization in a closed system, while the constant pressure autoclave method of Natta and Pasquon[2] (Fig. 1.2) and the use of a flask connected to an olefin gas storage unit by Keii *et al.*[6] (Fig. 1.3) and Ingberman *et al.*[7] (Fig. 1.4) are examples of open system techniques.

In the case of a closed system, reaction rates may be followed by measuring the change in the concentration of monomer with time, or by measuring the corresponding volume changes of the reaction mixture. The rate of polymer formation, *R*, can be calculated, in the absence of

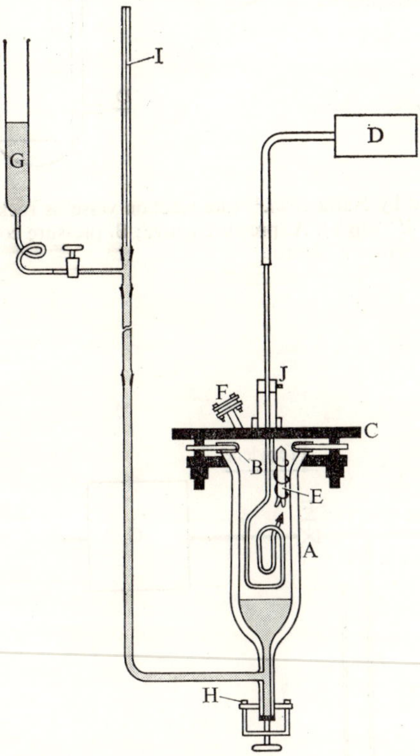

Fig. 1.1. Polymerization dilatometer used by Cooper *et al.*[4] A, 1/2 in. tube; B, seal with Teflon insert; C, mild steel plate; D, motor with flexible drive shaft; E, ampoule (VCl_3); F, aperture for insertion of syringe; G, mercury reservoir; H, level adjuster; I, mercury level; J, rotary seal for 200 rpm stainless steel stirrer rod. Total volume of reactant: 175 ml.

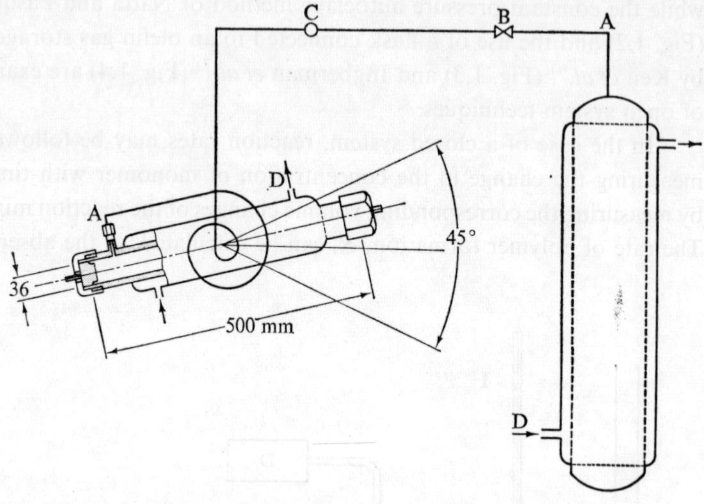

Fig. 1.2. Apparatus used by Natta *et al.*[2] (the reaction vessel is rocked 45 times per min through a 45° angle). A, pressure gauge; B, pressure control; C, flow indicator; D, temperature control.

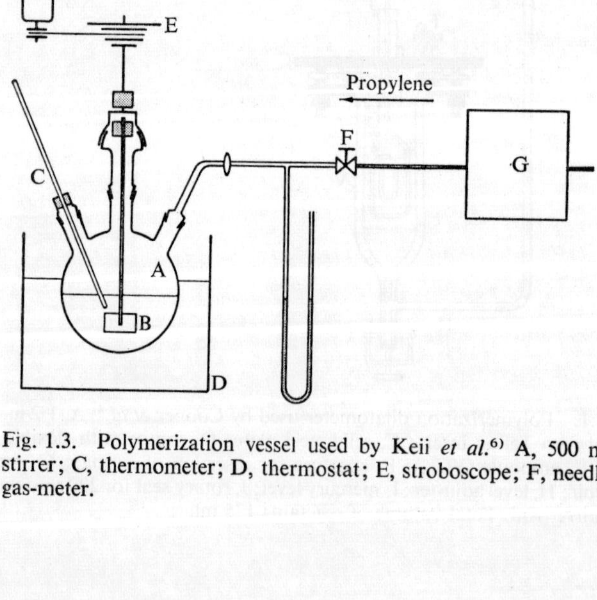

Fig. 1.3. Polymerization vessel used by Keii *et al.*[6] A, 500 ml flask; B, stirrer; C, thermometer; D, thermostat; E, stroboscope; F, needle valve; G, gas-meter.

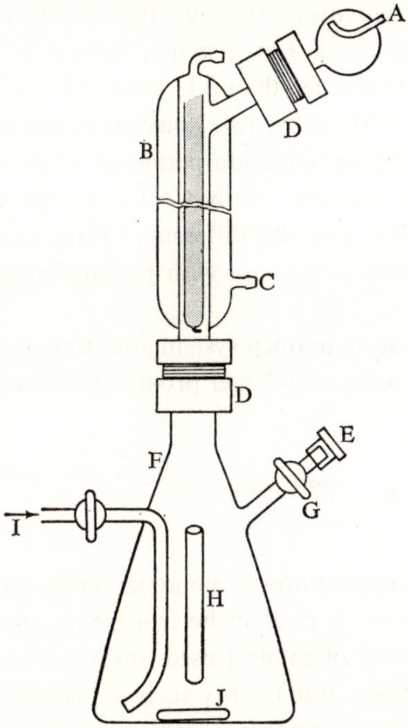

Fig. 1.4. Polymerization flask used by Ingberman *et al.*[7] A, exit to pressure gauge and flow meter; B, condenser; C, water; D, couplings for 1 in. Pyrex pipe flanges; E, injection port; F, modified 2 l flask; G, stopcocks; H, thermometer well; I, gas inlet; J, magnetic stirrer bar.

side reactions such as alkylation of the solvent by the monomer, from Eq. 1.1,

$$R = \frac{dm}{dt} = V\left(-\frac{d[\mathrm{M}]}{dt}\right) \tag{1.1}$$

where m is the mass of polymer formed by time t, V is the volume of the reaction mixture and [M] the monomer concentration. In closed systems, the "conversion factor" Q is often used, and this is defined as

$$Q = \frac{[\mathrm{M}]_0 - [\mathrm{M}]}{[\mathrm{M}]_0} \tag{1.2}$$

where $[\mathrm{M}]_0$ is the initial concentration of monomer. Many workers use the value of Q after polymerization for some fixed time, e.g. one hour, for the comparison of catalyst activities. However, the use of integral rates of this type is not always valid. For instance, the use of Q is significant only with values of Q smaller than 0.1, where $Q \propto t$,† and in general the differential rate ($d[\mathrm{M}]/dt$) or rate constant is preferable. In particular, for reactions having an induction period or where deactivation takes place during the reaction, catalyst activities cannot be compared by their integral rates, and the differential rates must be used. It should be noted that this is the case with polymerization by Ziegler-Natta catalysts.

The rate of polymerization R in an open system after time t can be defined as the rate at which the mass of polymer produced is increasing at that time.

$$R = \frac{dm}{dt} \tag{1.3}$$

This will be equal to the rate of consumption of monomer, in the absence of side reactions. Stationary reaction rates in heterogeneous catalysis should be proportional to the mass of catalyst used, since the reaction takes place on the catalyst surface, but factors such as temperature, monomer concentration (or partial pressure) and concentration of cocatalyst also influence the rate. The dependence of the reaction rate R on the polymerization time t during the induction or deactivation periods may be expressed as a kinetic curve of functional form

$$R = R(t) \tag{1.4}$$

The total number of moles of cocatalyst in a Ziegler-Natta catalyst is an extensive quantity, but since the effect of a solute species on the rate of reaction is dependent on the intensive quantity, i.e. the concentration or activity, it is not sufficient to specify the mass of cocatalyst only in any system. Similarly, the molar ratio of cocatalyst to catalyst,

† An n-th order rate equation at low levels of conversion ($Q \ll 1$) can be approximated as follows:

$$-\frac{d[\mathrm{M}]}{dt} = k[\mathrm{M}]^n$$

$$\therefore \quad \frac{dQ}{dt} = \frac{k}{[\mathrm{M}]_0^{\,1-n}}(1 - Q)^n \simeq \frac{k}{[\mathrm{M}]_0^{\,1-n}}$$

Thus the conversion, Q, at a given time can in this case be taken as a measure of the rate constant, or activity.

e.g. [A]/[Ti] for an $AlEt_3$ and $TiCl_3$ system, is not sufficient unless the volume of the reaction mixture is also given.

A linear dependence of the stationary reaction rate on the surface area or mass of the catalyst provides strong evidence of kinetic control, indicating in general that the reaction occurs only on the catalyst surface. However, in gas-liquid-solid systems such as many Ziegler-Natta catalysts, physical processes such as absorption of gas by the liquid phase, or diffusion of the solute into the surface of the solid phase may take place. This may affect the rate of reaction and in such cases, where the reaction is under partial kinetic control and partial physical control, the linear relationship between the reaction rate and the mass of catalyst will no longer hold. This situation will be discussed in the following section.

1.3 PHYSICAL PROCESSES IN ZIEGLER-NATTA POLYMERIZATION

The polymerization of a gaseous monomer in an open system may require an effective mass transfer system to avoid the formation of concentration gradients of monomer in the solution; this is usually done by mechanical agitation of the solution. Kinetic measurements are generally made with the gas and liquid phases in equilibrium, and it is often convenient to carry out these measurements at relatively low temperatures and reaction rates, where the proportionality between the reaction rate and the mass of catalyst is maintained, and where the observed activation energy may be as high as that of a chemical reaction. Thus, since the early work of Natta and Pasquon,[2)] who found both a linear relation between the reaction rate and the mass of catalyst, and a high energy of activation (10 kcal/mole) for the polymerization of propylene with a $TiCl_3/AlEt_3$ catalyst, there has been little attempt to consider the role of physical processes in these reactions.

The importance of propylene absorption by the solute in the same system was first pointed out by Keii *et al.*,[6)] using the constant pressure reactor shown in Fig. 1.3. The catalyst system used was $TiCl_3$ (AA-grade, Stauffer Chemical Co.)–$AlEt_3$–*n*-heptane. Fig. 1.5 shows the kinetic curves obtained at various stirring speeds, and illustrates transition phenomena due to sudden changes in stirring speed. These may be accounted for as follows. Let the concentration of propylene monomer

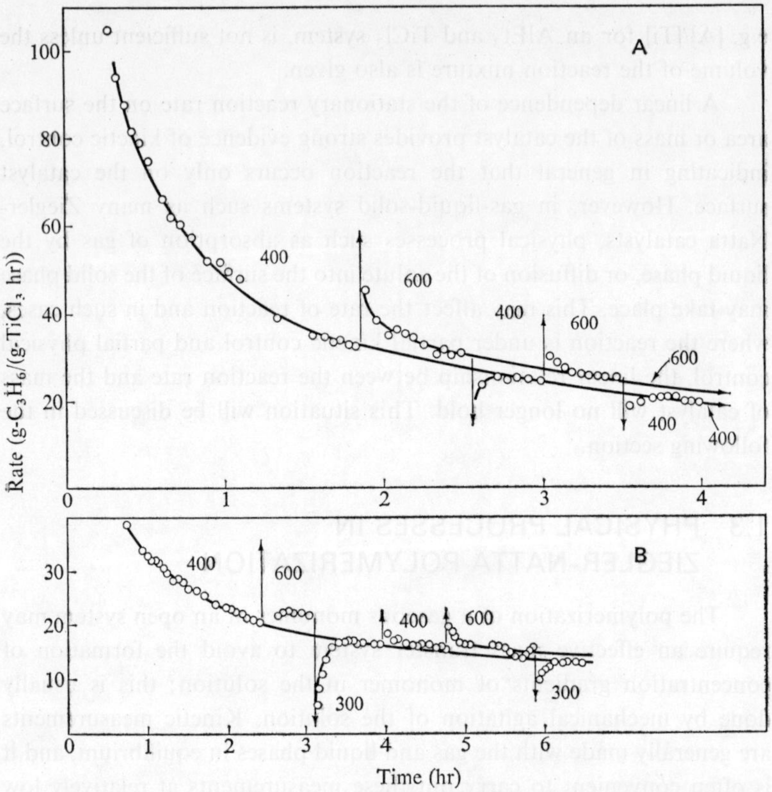

Fig. 1.5. Dependence of polymerization rate on stirring speed.[6] A, 0.505 g-$TiCl_3$ in 250 ml *n*-heptane at 44°C. $P = 650$ mmHg, [A]/[Ti] = 1.0. B, 0.480 g-$TiCl_3$ in 542 ml *n*-heptane at 27°C. $P = 710$ mmHg, [A]/[Ti] = 1.0.

in the solution be [M], and the monomer concentration at the gas-liquid interface, $[M]_0$ (assumed to be in equilibrium with the gaseous propylene). If k represents the polymerization rate constant and β the mass transfer coefficient of propylene, then the rate of change of propylene concentration in the bulk solution at time t may be expressed as

$$\frac{d[M]}{dt} = \beta([M]_0 - [M]) - kG[M] \tag{1.5}$$

The first term on the right-hand side represents the supply of fresh monomer and the second gives the consumption of monomer per unit time and volume due to polymerization by a catalyst of mass G. Assuming that the dependence of polymerization rate on time at a given stirring speed can be neglected in comparison with that found on transition between different speeds,[†] a stationary state equation, with $d[\mathrm{M}]/dt = 0$, can be applied to each kinetic curve. Thus at a stirring speed a we have

$$R_a = \beta_a([\mathrm{M}]_0 - [\mathrm{M}]_a) = kG[\mathrm{M}]_a \tag{1.6a}$$

and at a different stirring speed, b,

$$R_b = \beta_b([\mathrm{M}]_0 - [\mathrm{M}]_b) = kG[\mathrm{M}]_b \tag{1.6b}$$

where R is the rate of polymerization. Eq. 1.6*a* and 1.6*b* may be expressed as first order rate equations with respect to $[\mathrm{M}]_0$ as follows.

$$R_a = \frac{\beta_a kG}{\beta_a + kG}[\mathrm{M}]_0 \tag{1.7a}$$

$$R_b = \frac{\beta_b kG}{\beta_b + kG}[\mathrm{M}]_0 \tag{1.7b}$$

When $\beta \gg kG$, the reaction will be kinetically controlled and the rate equation reduces to

$$R = kG[\mathrm{M}]_0 \tag{1.8a}$$

but where $\beta \ll kG$ the rate of polymerization is determined by the diffusion process, as indicated in Eq. 1.8*b*.

$$R = \beta[\mathrm{M}]_0 \tag{1.8b}$$

We shall now consider the effect of a sudden change in stirring speed from a to b ($a > b$). Eq. 1.5 may be applied to the transition state as follows.

$$\frac{d[\mathrm{M}]}{dt} = \beta_b([\mathrm{M}]_0 - [\mathrm{M}]) - kG[\mathrm{M}]$$

with $$[\mathrm{M}] = [\mathrm{M}]_a = \frac{\beta_a}{\beta_a + kG}[\mathrm{M}]_0 \quad \text{at} \quad t = 0$$

† The relaxation times of the kinetic curves are 1.5–2.0 hr, whereas those of the transition curves are < 10 min.

$$[\text{M}] = [\text{M}]_b = \frac{\beta_b}{\beta_b + kG}[\text{M}]_0 \quad \text{at} \quad t = \infty \tag{1.9}$$

Thus the increase in the propylene absorption rate at the time of the change in the stirring speed (i.e. $t = 0$) can be calculated,

$$\left(\frac{d[\text{M}]}{dt}\right)_0 = \beta_b([\text{M}]_0 - [\text{M}]_a) - kG[\text{M}]_a$$

$$= (\beta_b - \beta_a)([\text{M}]_0 - [\text{M}]_a) > 0 \tag{1.9a}$$

and similarly the increase in the absorption rate at $t = \infty$ will be

$$\left(\frac{d[\text{M}]}{dt}\right)_\infty = \beta_b([\text{M}]_0 - [\text{M}]_b) - kG[\text{M}]_b$$

$$= 0 \tag{1.9b}$$

The monomer concentration [M] at time t can be derived from Eq. 1.9,

$$[\text{M}] = [\text{M}]_a\, e^{-\alpha t} + [\text{M}]_b(1 - e^{-\alpha t})$$

where $\alpha = \beta_b + kG$ (1.9c)

and the change in the absorption rate during this transition state can now be obtained from Eq. 1.9 and 1.9c.

$$\frac{d[\text{M}]}{dt} = \alpha([\text{M}]_b - [\text{M}]_a)e^{-\alpha t} \tag{1.9d}$$

The form of Eq. 1.9d indicates that the increment of the absorption rate decreases exponentially to zero from a maximum value corresponding to Eq. 1.9b. In the case of a reduction in the stirring speed (e.g. from $b \to a$), the change in the rate of absorption will be negative. The use of the stirring speed jump technique can thus demonstrate the effect of mass transfer on the overall polymerization rate, though under normal polymerization conditions the effect is not very marked. Further investigations of the mass transfer effect were carried out by Keii *et al.*[8] using the reactor illustrated in Fig. 1.6. It was reported that the rate of polymerization was proportional to the stirring speed n.

$$R_n \propto n \tag{1.10}$$

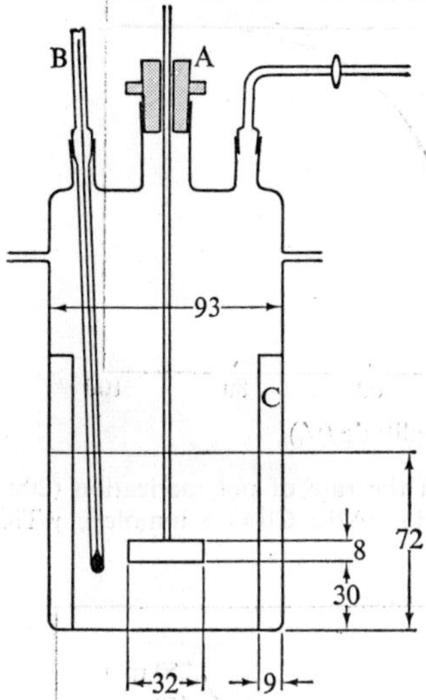

Fig. 1.6. Reactor used by Keii *et al.*[8] for the investigation of diffusional kinetics. A, teflon diaphragm; B, thermometer; C, baffle.

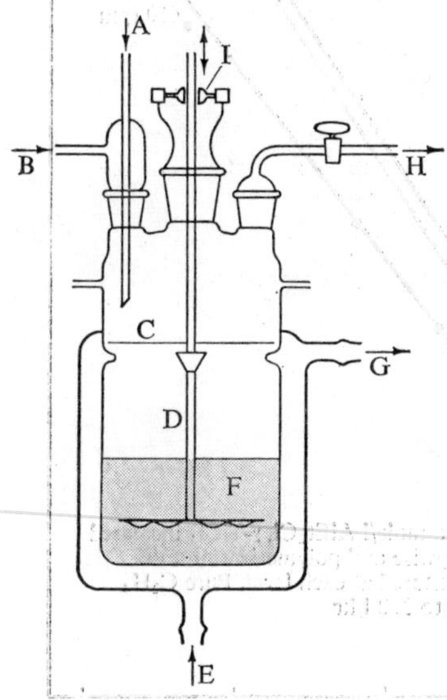

Fig. 1.7. Polymerization reactor used by Berger and Grieveson.[9] A, liquid feed streams; B, gas feed streams; C, stainless steel lid; D, stirrer; E, thermostatting liquid in; F, reaction volume (200 ml); G, liquid exit; H, vent; I, Neoprene diaphragm.

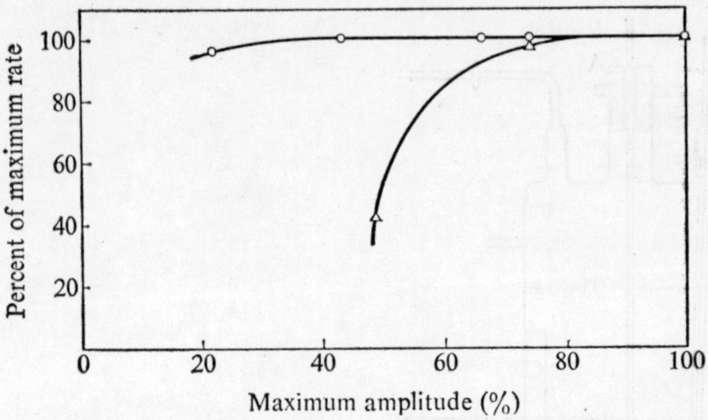

Fig. 1.8. Effect of stirring rate on the rate of polymerization (200 ml cyclohexane at 40°C, $P = 760$ mmHg, $[AlEt_2Cl] = 5$ mmole/l, γ-$TiCl_3 = 1$ mmole/l (○) or 6 mmole/l (△)).[9]

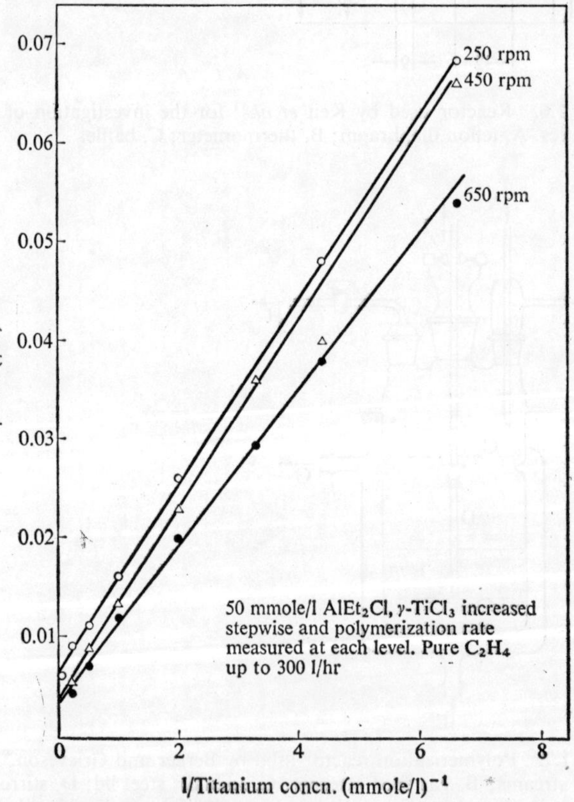

Fig. 1.9. Effect of mass transfer at different stirring rates on the rate of polymerization.[10]

The effect of agitation on ethylene polymerization using $TiCl_3$ and $AlEt_2Cl$ has been investigated by Berger and Grieveson.[9] Their reactor, using a vibrating stirrer, is illustrated in Fig. 1.7 and the results presented in Fig. 1.8. Boocock and Haward[10] also studied the effect of mass transfer processes on the polymerization of ethylene. Eq. 1.7*a* can be rewritten as

$$\frac{1}{R} = \frac{1}{kG[M]_0} + \frac{1}{\beta[M]_0} \tag{1.11}$$

These authors investigated ethylene polymerization with different amounts of $TiCl_3$ and at various stirrer speeds, and plotted the reciprocal of the polymerization rate against the reciprocal of the $TiCl_3$ concentration. The results are shown in Fig. 1.9. As can be seen from Eq. 1.11, the intercept gives the reciprocal of the maximum rate of solution of ethylene in the diluent, $\beta[M]_0$, and the slope gives the kinetic rate of polymerization $k[M]_0$. Fig. 1.10 shows the relationship between

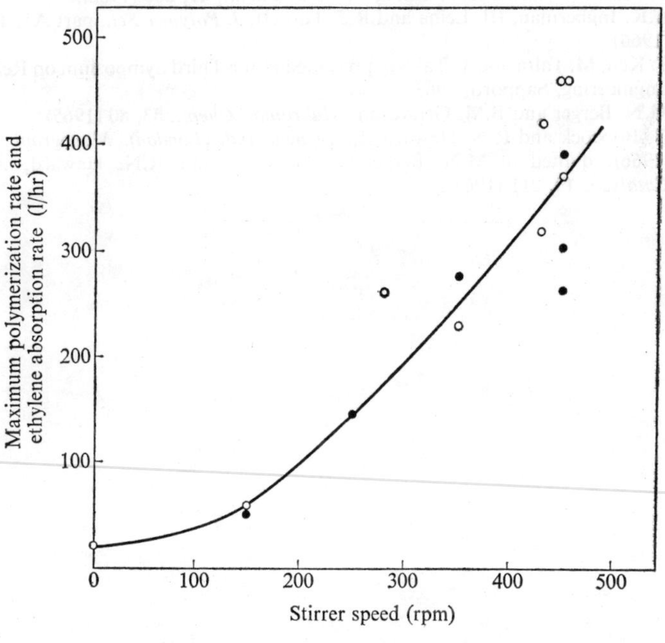

Fig. 1.10. Dependence of maximum polymerization rate (●) and maximum initial rate of solution (○) on the stirring rate.[10] (50 mmole/l $AlEt_2Cl$ and 50 mmole/l $TiCl_3$. Pure ethylene.)

the initial rate of solution of ethylene and maximum rate of polymerization, and the stirrer speed. The dependence appears linear at higher stirring speeds, and slightly higher than first order at lower speeds.

The effects of mass transfer may be even more significant in the polymerization of ethylene with $TiCl_3$ and $AlEt_3$, where the rate constant k is larger than for the $TiCl_3/AlEt_2Cl$ catalysts used above.

References

1. D.O. Jordan, *The Stereochemistry of Macromolecules* (ed. A.D. Ketley), vol. 1, p. 3, Dekker, 1967.
2. G. Natta and I. Pasquon, *Advan. Catalysis*, **11**, 1 (1959).
3. J. Ambroz, P. Osecky, J. Mejzlik and O. Hamrik, *J. Polymer Sci.*, **C16**, 423 (1967).
4. W. Cooper, D.E. Evans, G.D.T. Owen and G. Vaughan, *J. Polymer Sci.*, **C4**, 211 (1964).
5. K. Fukui, T. Kagiya, S. Machi, T. Shimizu and S. Yuasa, *Bull. Chem. Soc. Japan*, **35**, 303, 306 (1962).
6. T. Keii, M. Taira and T. Takagi, *Can. J. Chem.*, **41**, 206 (1962).
7. A.K. Ingberman, I.J. Leina and R.J. Turbett, *J. Polymer Sci.*, part A1, **4**, 2781 (1966).
8. T. Keii, M. Taira and T. Takagi, presented at the Third Symposium on Reaction Engineering, Sapporo, 1963.
9. M.N. Berger and B.M. Grieveson, *Makromol. Chem.*, **83**, 80 (1965).
10. G. Boocock and R.N. Haward, *Soc. Chem. Ind. (London), Monograph*, **20**, 3 (1966); quoted in M.N. Berger, G. Boocock and R.N. Haward, *Advan. Catalysis*, **19**, 211 (1968).

Chapter 2

Kinetics of propylene polymerization

2.1 GENERAL KINETIC BEHAVIOR OF $TiCl_3$ AND ORGANOMETALS

The polymerization rate, *R*, of propylene under constant pressure using an active† $TiCl_3/AlEt_3$ catalyst changes markedly with time, as illustrated by curve A in Fig. 2.1.[1)] If active $TiCl_3/AlEt_2Cl$ or non-active† $TiCl_3/AlEt_3$ catalyst systems are used, the change is not so significant, as shown by curve B in the same figure. These kinetic curves may be divided into three parts in the case of curve A and two parts for curve B. Curve A shows an induction or build-up period (I), where the rate of polymerization increases rapidly to a maximum followed by a decay period (II) during which the rate decreases gradually to the stationary state (III). In the case of curve B there is a gradual increase in the rate of polymerization in the build-up period (I) until the stationary state (III) is reached directly. Curves of type A will be referred to as "decay" type and those of type B, as "acceleration", or "build-up" type.

The kinetic curves obtained during propylene polymerization by active $TiCl_3/ZnEt_2$[2)] and $VOCl_3/AlEt_3$[3)] catalyst systems are of the decay type, whereas all curves obtained with $TiCl_3/AlEt_2Cl$ can be classified as acceleration type, though they sometimes show a small maximum. Kinetic curves obtained during ethylene polymerization with $TiCl_3$ and $AlEt_3$ are of the decay type, but the build-up period

† "Active" $TiCl_3$ refers to fresh α, γ or δ-$TiCl_3$ or to commercial active catalysts such as AA- and HA-grades (Stauffer Chemical Co.) or TAC-grade (Tohō Titanium Co.). "Non-active" refers to old or deactivated samples of $TiCl_3$, or to non-activated commercial grades, e.g. A-, H- and TA-grades.

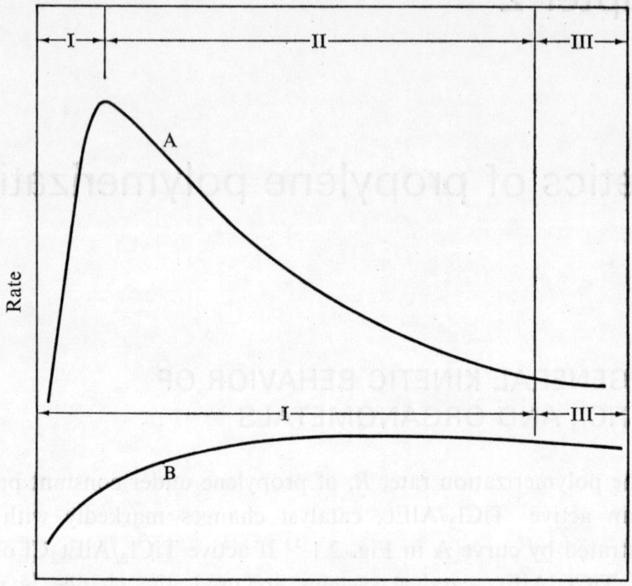

Fig. 2.1. Typical kinetic curves obtained during propylene polymerization with $TiCl_3$. A, decay type; B, build-up or acceleration type; I build-up period; II, decay period; III, stationary period.

is short because of the very rapid increase in reaction rate and the existence of period (I) may not be immediately apparent. Since the duration of periods (I) and (II) is typically about two hours under normal propylene polymerization conditions using $TiCl_3$ and an organometal compound, it can readily be seen that the use of integral polymerization rates, or the conversion factor, Q, as a measure of catalyst activity is unlikely to prove helpful.

The existence of the two types of kinetic curve was noted by Natta *et al.*[4] who proposed an interesting explanation. They found that the kinetic curves obtained with unground samples of α-$TiCl_3$ were of the acceleration type whereas those obtained with ground samples of the same $TiCl_3$ were of the decay type, though in both cases the stationary rate was independent of the initial state of the $TiCl_3$. On the basis of these results they suggested that the size of both the large unground particles and the small ground particles was changed during the reaction to a common size appropriate for the particular polymerization conditions. This cleavage and agglomeration of particles was thought to arise from the mechanical action of polymer chains formed on the catalyst surfaces, and gave rise to their use of the term "adjustment period" for regions (I) and (II) of the kinetic curves shown in Fig. 2.1.

2.2 THE BUILD-UP PERIOD

2.2.1 The build-up period in acceleration type polymerization

The acceleration type kinetic curves obtained during propylene polymerization with unground samples of α-$TiCl_3$ and $AlEt_3$ were studied by Natta *et al.*,[5,6] using the constant pressure, open system illustrated in Fig. 1.2. Curves were obtained (Fig. 2.2 and 2.3) at various temperatures and propylene pressures, using the same amount of α-$TiCl_3$ catalyst in each case, and were characterized by measuring the time, $t_{3/4}$, required for the polymerization rate to reach a value equal to three quarters that of the stationary rate R_∞. The values obtained for $t_{3/4}$ were found to depend strongly on the stationary rate, as shown in Fig. 2.4. These results were considered by Natta *et al.*[4] to support their suggestions concerning catalyst particle size. Keii[7,8] discussed this work together with experimental results obtained on decay type

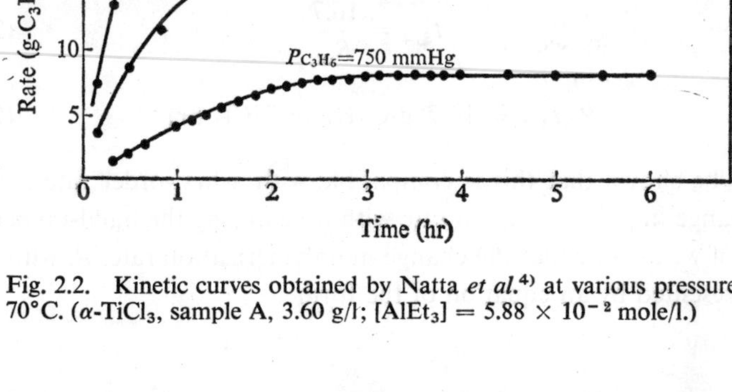

Fig. 2.2. Kinetic curves obtained by Natta *et al.*[4] at various pressures and 70°C. (α-$TiCl_3$, sample A, 3.60 g/l; $[AlEt_3] = 5.88 \times 10^{-2}$ mole/l.)

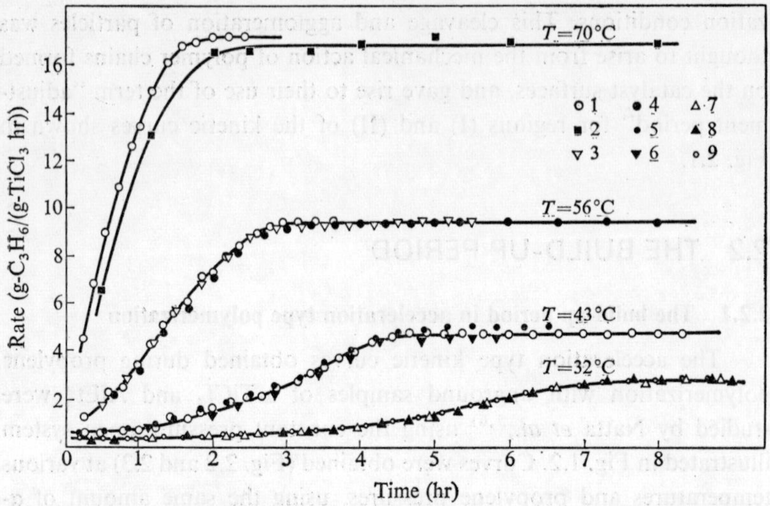

Fig. 2.3. Kinetic curves obtained by Natta *et al.*[4] at various temperatures with P = 1500 mmHg.

	α-$TiCl_3$ (sample A) g/l	$[Al(C_2H_5)_3]$ mole/l	[A]/[Ti] mole/mole
1	3.80	2.95×10^{-2}	1.18
2	0.80	4.45×10^{-2}	8.50
3	7.60	7.36×10^{-2}	1.50
4	1.60	5.90×10^{-2}	5.60
5	7.60	17.70×10^{-2}	3.60
6	10.80	7.36×10^{-2}	1.05
7	20.20	7.36×10^{-2}	0.52
8	12.16	17.70×10^{-2}	2.20
9	12.10	7.36×10^{-2}	0.94

kinetic curves.

The relationship shown in Fig. 2.4 between $t_{3/4}$ and $1/R_\infty$ may be expressed as

$$t_{3/4} = \frac{16.7}{R_\infty} \tag{2.1a}$$

or

$$R_\infty t_{3/4} = 16.7 \text{ g-C}_3\text{H}_6/(\text{g-TiCl}_3 \text{ hr}) \tag{2.1b}$$

It can be shown that this is compatible with a first-order rate law for the change in polymerization rate with time during the build-up period. Thus, if we assume that the change in polymerization rate, R, with time is represented by an equation of the form

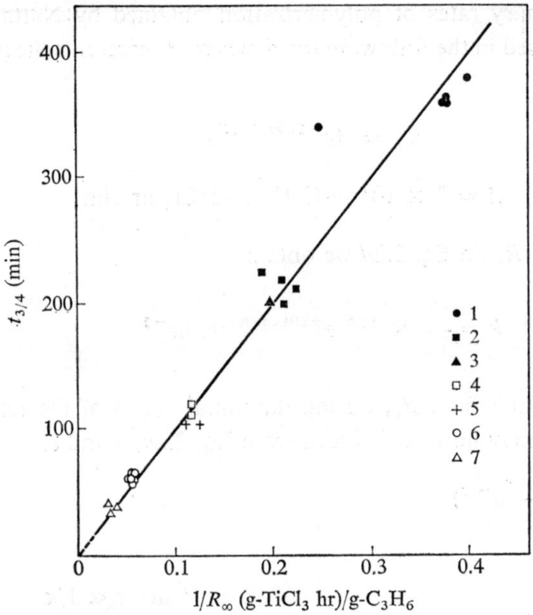

Fig. 2.4. Relationship between $t_{3/4}$ and the reciprocal of the stationary rate of polymerization.[4] 1, 32°C and 1680 mmHg; 2, 43°C and 1640 mmHg; 3, 32°C and 2680 mmHg; 4, 56°C and 1570 mmHg; 5, 70°C and 750 mmHg; 6, 70°C and 1450 mmHg; 7, 70°C and 2450 mmHg.

$$\frac{dR}{dt} = k(R_\infty - R) \tag{2.2a}$$

Eq. 2.2*b* is obtained on integration.

$$R = R_\infty(1 - e^{-kt}) \tag{2.2b}$$

Putting $R/R_\infty = 3/4$ at $t = t_{3/4}$,

$$t_{3/4} = \frac{1.39}{k} \tag{2.2c}$$

Comparing this with the experimental result, Eq. 2.2*a*, it is apparent that

$$k = \frac{R_\infty}{12} \quad \left(\frac{\text{g-C}_3\text{H}_6}{\text{g-TiCl}_3\ \text{hr atm}}\right) \Big/ \left(\frac{\text{g-C}_3\text{H}_6}{\text{g-TiCl}_3\ \text{atm}}\right) \tag{2.2d}$$

The stationary rates of polymerization obtained by Natta *et al.* could be expressed in the following form, where P represents the propylene pressure.

$$R_\infty = Ae^{-10\text{kcal}/RT} P, \tag{2.3}$$

where $$A = 2 \times 10^7 \text{ g-}C_3H_6/(\text{g-}TiCl_3 \text{ hr atm})$$

Substituting for R_∞ in Eq. 2.2*d* we obtain

$$k = \frac{2}{1.2} \times 10^6 \, e^{-10\text{kcal}/RT} P \text{ hr}^{-1} \tag{2.2e}$$

The polymerization rate, R_0, during the initial stages of the reaction, where $t \ll 1/k$, may now be deduced from Eq. 2.2*b*, *d* and *e*.

$$\begin{aligned} R_0 &= R_\infty(1 - e^{-kt}) \\ &\simeq R_\infty kt \\ &= \frac{1}{12} R_\infty^2 t \qquad \text{at} \quad t \ll 1/k \qquad (2.4) \\ &= 3.3 \times 10^{13} \, e^{-20/RT} P^2 t \text{ g-}C_3H_6/(\text{g-}TiCl_3 \text{ hr}) \end{aligned}$$

It has been found that Eq. 2.4 can also be applied to the initial rate during the build-up period of decay type kinetic curves, though the value of the pre-exponential factor will then be different.

2.2.2 The build-up period in decay type polymerization

The kinetics of the build-up period during decay type polymerization have been studied by Keii, Soga *et al.*,[9] using the reactor shown in Fig. 1.3. They found that the kinetic behavior of propylene polymerization during the build-up period was strongly affected by the order of addition of $AlEt_3$ and C_3H_6 to the $TiCl_3$. The results will be described below.

Case A $(TiCl_3 + AlEt_3) + C_3H_6$

A fixed amount of $AlEt_3$ was syringed into a reactor containing 250 ml of *n*-heptane and 0.5 g of $TiCl_3$ (AA-grade, Stauffer Chemical Co. Specific surface area, 25 m^2/g). The mixture was agitated under nitrogen for ten minutes at the polymerization temperature, then the nitrogen was pumped out and propylene introduced. Although the initial polymerization rate was disturbed by the solution of propylene into the solvent, it could be measured after about five minutes. Fig. 2.5 shows the kinetic curves obtained at various propylene pressures. The initial rates of polymerization, R_0, can in this case be regarded as proportional to the time of polymerization. From Fig. 2.6, which shows the dependence of the slope, R_0/t, on the propylene pressure, P, it can be seen that

$$R_0 \propto P^2 t \tag{2.5a}$$

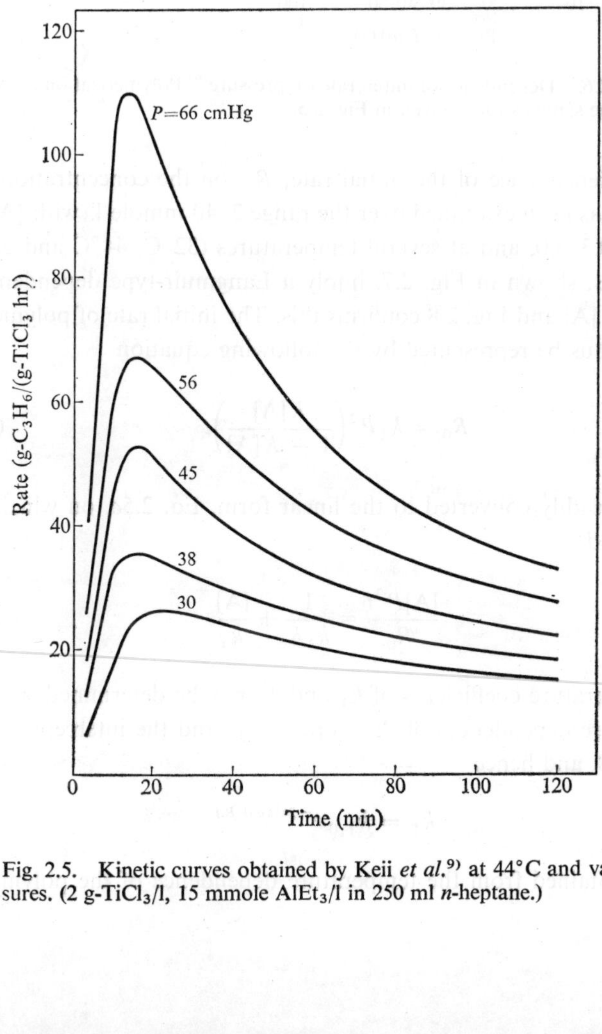

Fig. 2.5. Kinetic curves obtained by Keii *et al.*[9] at 44°C and various pressures. (2 g-$TiCl_3$/l, 15 mmole $AlEt_3$/l in 250 ml *n*-heptane.)

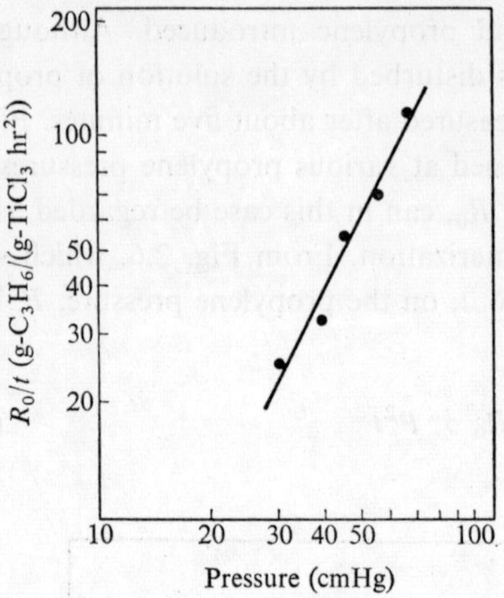

Fig. 2.6. Dependence of initial rate on pressure.[9] Polymerization conditions are the same as those given in Fig. 2.5.

The dependence of the initial rate, R_0, on the concentration, [A], of $AlEt_3$ has been examined over the range 2–40 mmole/l, with [A]/[Ti] ratios of 0.3–3.0, and at several temperatures (32°C, 44°C and 57°C). The results, shown in Fig. 2.7, imply a Langmuir-type dependence of R_0/P^2t on [A] and Fig. 2.8 confirms this. The initial rate of polymerization can thus be represented by the following equation

$$R_0 = k_1 P^2\left(\frac{K[A]}{1 + K[A]}\right)t \tag{2.5b}$$

which is readily converted to the linear form, Eq. 2.5*c*, on which Fig. 2.8 is based

$$\frac{[A]P^2t}{R_0} = \frac{1}{k_1 K} + \frac{[A]}{k_1} \tag{2.5c}$$

The temperature coefficients of k_1 and K may be determined from the temperature dependences of the slope ($1/k_1$) and the intercept ($1/k_1K$) in Fig. 2.8, and hence

$$k_1 = k_{1,0}e^{-20\text{kcal}/RT} \tag{2.5d}$$

This is obtained from the temperature dependence of the polymeriza-

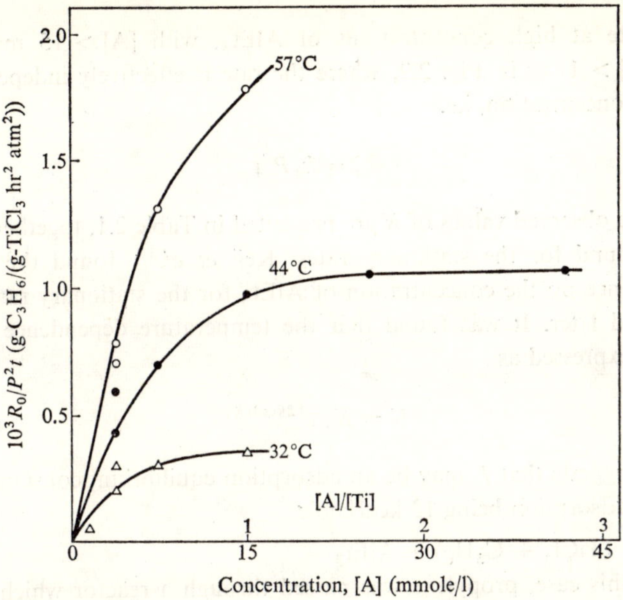

Fig. 2.7. Dependence of initial rate on concentration of $AlEt_3$ at various temperatures.[9] ([$TiCl_3$] = 2 g/l in 250 ml *n*-heptane.)

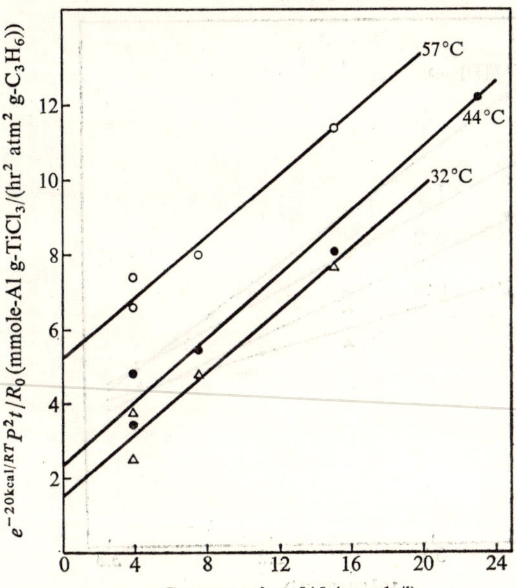

Fig. 2.8. Langmuir dependence of initial rate on $AlEt_3$ concentration.[9]

tion rate at high concentrations of $AlEt_3$, with [A] > 15 mmole/l ([A]/[Ti] > 1) as in Fig. 2.9, where the rate is effectively independent of the concentration, i.e.

$$R_0 = k_1 P^2 t \tag{2.5e}$$

The observed values of K are presented in Table 2.1, together with those found for the stationary rates. Keii *et al.*[10] found the same dependence on the concentration of $AlEt_3$ for the stationary rates, as discussed later. It was found that the temperature dependence of K can be expressed as

$$K = K_0\, e^{12\mathrm{kcal}/RT} \tag{2.5f}$$

which suggests that K may be an adsorption equilibrium constant, the heat of adsorption being 12 kcal/mole.

Case B ($TiCl_3 + C_3H_6$) + $AlEt_3$

In this case, propylene was passed through a reactor which contained 250 ml of *n*-heptane and 0.5 g of $TiCl_3$. After agitation for ten

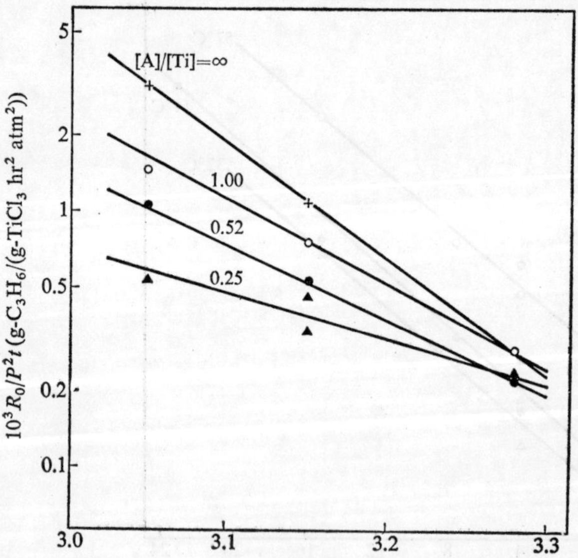

Fig. 2.9. R_0/P^2t *vs.* temperature.[9] ([$TiCl_3$] = 2 g/l).

TABLE 2.1 Values of adsorption constant, K, obtained from initial rate and stationary rate†

Temp. (°C)	K (l/mole) from initial rate	K (l/mole) from stationary rate	E (kcal/mole)
32	280	220	
44	170	120	12
57	70∼80	60	

† $TiCl_3$ = 2 g/l, [A] = 2∼45 mmole/l.

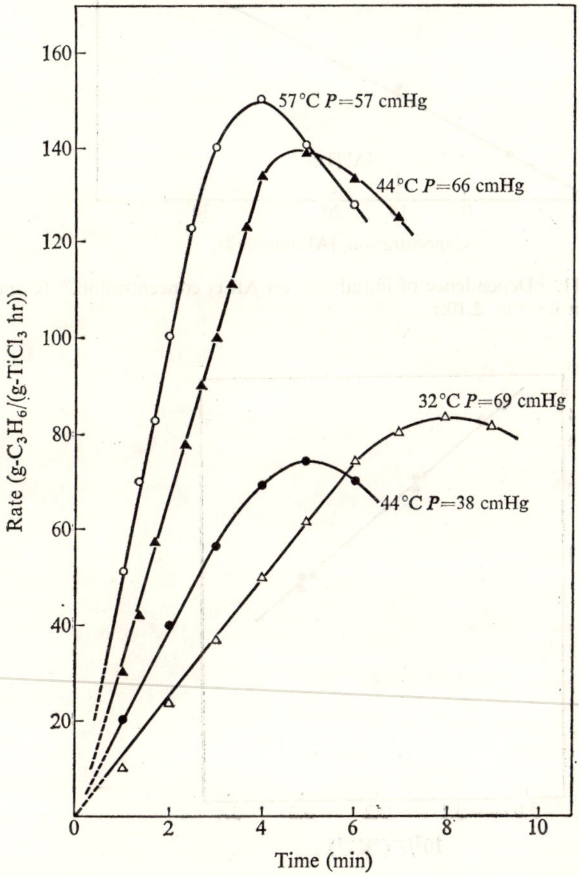

Fig. 2.10. Kinetic curves at various temperatures and pressure.[9] ([$TiCl_3$] = 2 g/l, [$AlEt_3$] = 15 mmole/l, 250 ml n-heptane).

minutes at the desired polymerization temperature, a fixed amount of $AlEt_3$ was syringed into the reactor. It proved possible to measure the rate of polymerization very soon after the addition of the $AlEt_3$, and some typical kinetic curves are shown in Fig. 2.10. Both the initial

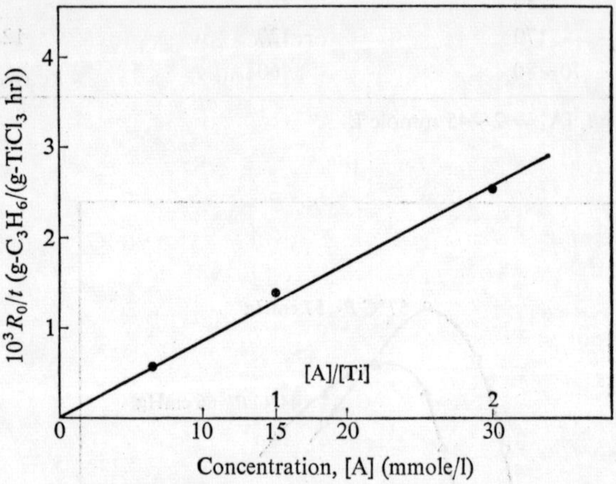

Fig. 2.11. Dependence of initial rate on $AlEt_3$ concentration.[9] (Conditions as given for Fig. 2.10.)

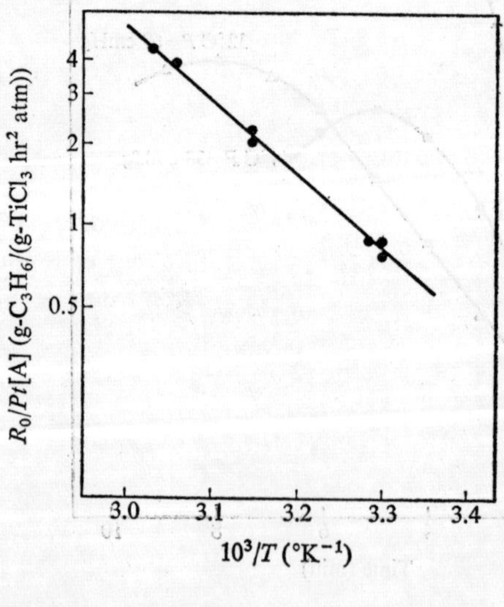

Fig. 2.12. Temperature dependence of initial rate.[9] (Conditions as in Fig. 2.10.)

and maximum rates of reaction were found to be greater than in case A as can be seen by comparison with Fig. 2.5. The dependence of the rate on propylene pressure and on $AlEt_3$ concentration is shown in Fig. 2.10 and 2.11 respectively, and from these results it can be shown that

$$R_0 = k_2 P[\mathrm{A}]t \tag{2.6a}$$

The activation energy of k_2 was determined by a plot of $R_0/P[\mathrm{A}]t$ against $1/T$ as shown in Fig. 2.12, and was found to be 17 kcal/mole. Hence we obtain the equation

$$k_2 = k_{2,0}e^{-17\mathrm{kcal}/RT} \tag{2.6b}$$

A theoretical discussion of these results by Keii *et al.* will be presented in the next chapter.

2.3 THE DECAY PERIOD

The decay period is that portion of the kinetic curve where the polymerization rate decreases gradually from its maximum value to the stationary value, as defined in (2.1), and forms the main part of decay type kinetic curves. Keii[7] proposed a kinetic equation for the reaction rate in this period for propylene polymerization with $TiCl_3$ and $AlEt_3$,

$$\frac{R - R_\infty}{\bar{R}_0 - R_\infty} = e^{-k_3 t} \tag{2.7a}$$

or

$$R = R_\infty + (\bar{R}_0 - R_\infty)e^{-k_3 t} \tag{2.7b}$$

where R_∞ is the stationary rate and k_3 and $\bar{R}_0$ are constants. This equation has been supported by Lefevre *et al.*,[11] who examined the polymerization of propylene with $TiCl_3/AlEt_3$ in an octane solvent, with the results shown in Fig. 2.13. Ambroz *et al.*[12] also proposed an equation of the same form as Eq. 2.7*a* and *b*.

Keii[7] noted that there is an analogy between Eq. 2.2*b* and 2.7*a* and *b*. On differentiating Eq. 2.7*b* with respect to *t*, we obtain

$$-\frac{dR}{dt} = k_3(R - R_\infty) \tag{2.7c}$$

which is of the same form as Eq. 2.2*a*, from which Eq. 2.2*b* is derived.

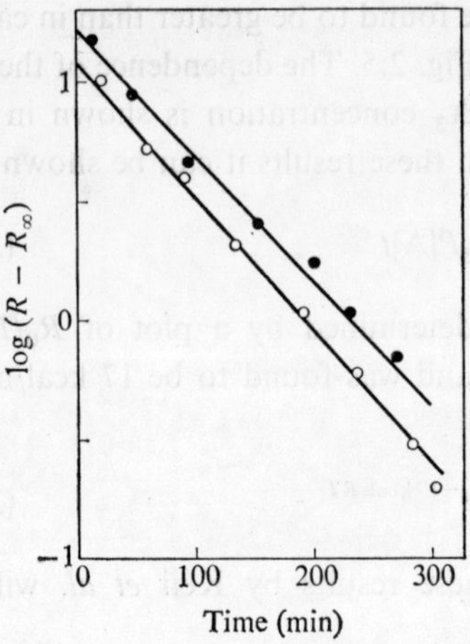

Fig. 2.13. Kinetic data for propylene polymerization according to Lefevre *et al.*[11] (●, $TiCl_3$ = 1.6 g/l, $AlEt_3$ = 7.5 g/l; ○, $TiCl_3$ = 1.2 g/l, [$AlEt_3$] = 4.7 g/l.)

Eq. 2.2*b* may thus be regarded as equivalent to Eq. 2.7*a* and *b* with $\bar{R}_0 = 0$, as in Eq. 2.8.

$$\frac{R_\infty - R}{R_\infty} = e^{-k_3 t} \tag{2.8}$$

Cooper *et al.*[13] applied Eq. 2.7*a* to the polymerization of isoprene by a $VCl_3/AlEt_3$ catalyst, using a dilatometer of the type illustrated in Fig. 1.1.

The constant k_3 may be considered as the constant of the first-order rate law governing the rate of approach of R to R_∞, and it is therefore important to clarify its behavior. This has been studied by Keii, Soga *et al.*[10] Fig. 2.14 shows the kinetic curves obtained with various concentrations of $AlEt_3$ (1.5 ~ 15 mmole/l) at 44 °C. Fig. 2.15 interprets these data in the light of Eq. 2.7*a*, and indicates that k_3 is independent of the $AlEt_3$ concentration, but Fig. 2.16 shows that k_3 is to some degree dependent on the propylene pressure. From the temperature dependence of the intercepts and slopes of the lines in Fig. 2.16, we obtain

$$k_3 = k'_{3,0} e^{-(2\sim 3)\text{kcal}/RT} + k''_{3,0} e^{-12\text{kcal}/RT} P \tag{2.9}$$

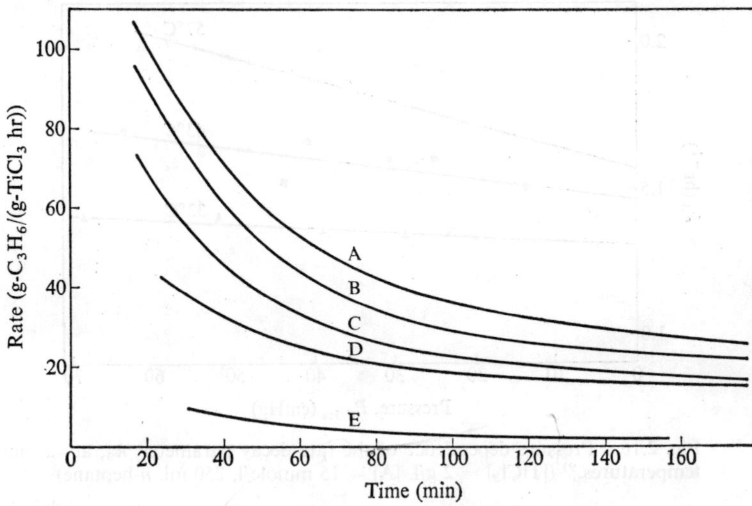

Fig. 2.14. Kinetic curves at various $AlEt_3$ concentrations.[9] ([$TiCl_3$ = 2 g/l in 250 ml *n*-heptane at 44°C. [A]/[Ti]; A = 1.0, B = 0.52, C = 0.41, D = 0.25, E = 0.11)

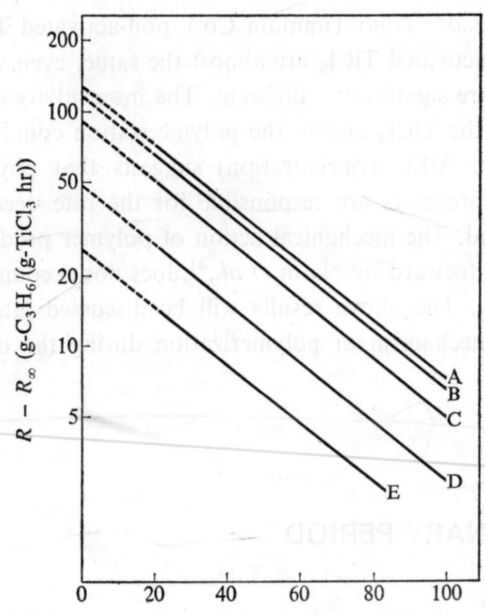

Fig. 2.15. Dependence of the kinetic data on time at various $AlEt_3$ concentrations.[9] (Conditions as given for Fig. 2.14)

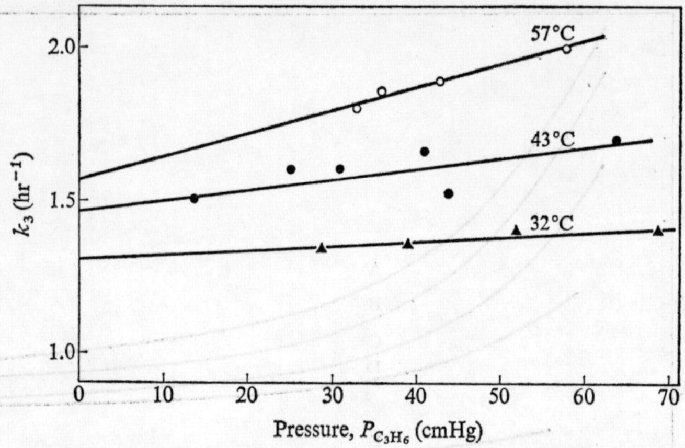

Fig. 2.16. Pressure dependence of the rate decay parameter, k_3, at various temperatures.[9] ($[TiCl_3]$ = 2 g/l, [A] = 15 mmole/l, 250 ml. *n*-heptane)

The results show that under normal polymerization conditions, the first term is larger than the second term. The values of k_3 obtained with various samples of $TiCl_3$, including active $TiCl_3$ (AA- and HA-Stauffer Chemical Co.; TAC-grade, Toho Titanium Co.), non-activated $TiCl_3$ (A-, H-grades) and deactivated $TiCl_3$ are almost the same, even when values of $\bar{R}_0$ and R_∞ are significantly different. The insensitivity of k_3 to both the nature of the $TiCl_3$ and to the polymerization conditions (temperature, pressure, $AlEt_3$ concentration) suggests that physical rather than chemical processes are responsible for the rate decrease during the decay period. The mechanical action of polymer produced on the catalyst, as put forward by Natta *et al.*,[4] does not account for this insensitivity in k_3. The above results will be discussed later in connection with the mechanism of polymerization during the decay period.

2.4 THE STATIONARY PERIOD

2.4.1 Natta's rate law

The rate law given in Eq. 2.3, proposed by Natta *et al.*,[4] has been supported by a number of authors, though it does not take into account the influence of alkylaluminium concentration. Natta *et al.*[4] summarized their own kinetic data on the basis of Eq. 2.3 or 2.10, in which G denotes the mass of $TiCl_3$.

$$R_\infty = Ae^{-10\text{kcal}/RT}GP \tag{2.10}$$

They found $A = 2 \times 10^7$ for unground α-$TiCl_3$ (sample A) and $A = 3.4 \times 10^7$ g-C_3H_6/(g-$TiCl_3$ hr atm) for ground samples of the same catalyst (sample B), though in the latter case the cocatalyst ($AlEt_3$) concentration was rather higher ($[A]/[Ti] \gtrsim 1$). As described in 2.1, they found acceleration type kinetic curves with unground α-$TiCl_3$ and decay type curves with ground α-$TiCl_3$. The experimental results on which the above rate law (Eq. 2.10) was based are shown in Fig. 2.3, 2.17, 2.18 and 2.19. It can be seen that this work was carried out at rather high propylene pressures (750–2700 mm Hg) and at high $AlEt_3$ cocatalyst concentrations (2.95–17.7×10^{-2} mole/l). Natta *et al.*[14] also examined the pressure dependence of the polymerization rate at lower propylene pressures, and obtained the results shown in Fig. 2.20. Although similar pressure dependence has sometimes been observed in the present author's laboratory, a linear relationship between the

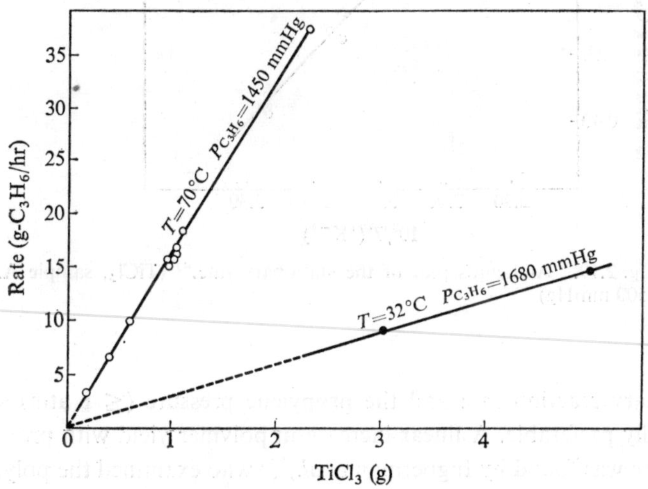

Fig. 2.17. Dependence of the stationary rate on mass of catalyst according to Natta *et al.*[4] ($TiCl_3$ sample A, 1450 mmHg, 70°C.)

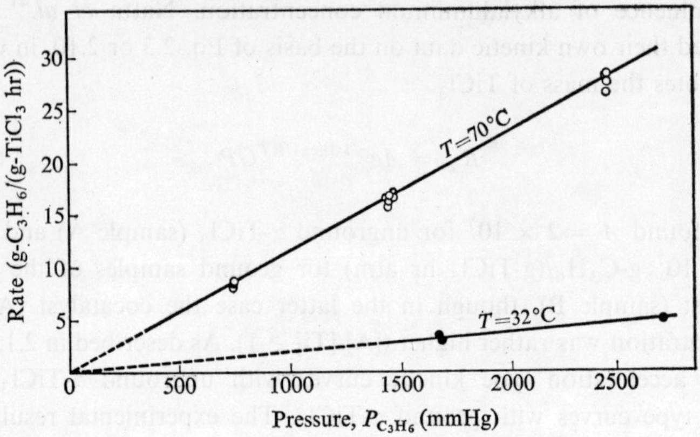

Fig. 2.18. Pressure dependence of the stationary rate.[4] (α-$TiCl_3$, sample A)

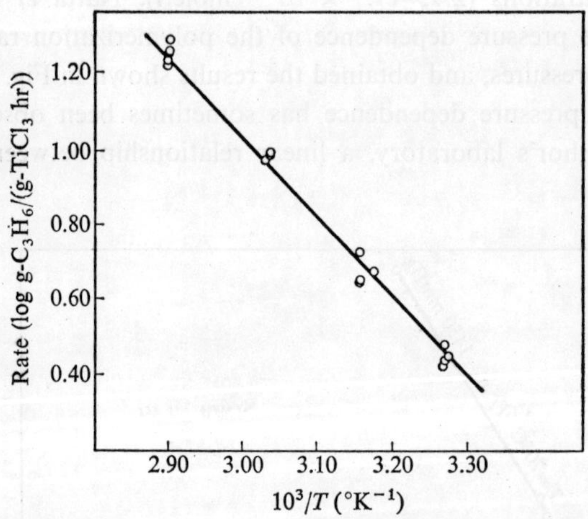

Fig. 2.19. Arrhenius plot of the stationary rate.[4] ($TiCl_3$, sample A. P = 1500 mmHg)

stationary reaction rate and the propylene pressure ($\leq$ 1 atm) seems generally preferable. A linear increase in polymer yield with propylene pressure was found by Ingberman *et al.*,[15] who examined the polymerization of propylene with α- and γ-$TiCl_3$ and $AlEt_2Cl$. It may be noted here that some degree of second-order pressure dependence has been

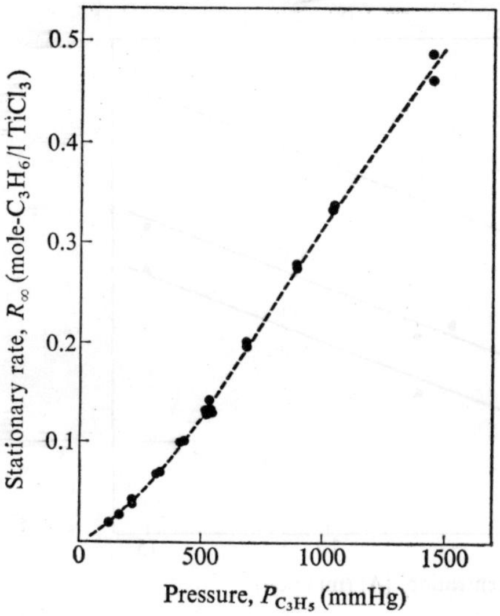

Fig. 2.20. Pressure dependence of the rate at low propylene pressure.[4] ($[TiCl_3]$ = 1.76 g/l, $[AlEt_3]$ = 0.06 mole/l, in toluene at 72°C)

clearly confirmed in the polymerization of ethylene with $TiCl_4$ and alkylaluminium.[16]

2.4.2 Dependence of polymerization rate on alkylaluminium concentration

Many investigations have been carried out to determine the influence of $AlEt_3$ or $AlEt_2Cl$ concentration on the rate of propylene polymerization. Schnecko *et al.*[17] and Keii, Soga *et al.*[9] independently obtained the same results, summarized in Eq. 2.11*a*.

$$R_\infty = Ae^{-10\text{kcal}/RT}GP\left(\frac{K[A]}{1 + K[A]}\right) \qquad (2.11a)$$

As shown in Fig. 2.21, Keii and Soga confirmed this dependence of the polymerization rate per unit mass of $TiCl_3$ on the concentration of $AlEt_3$, using the linear form of Eq. 2.11*a*.

$$\frac{[A]}{R_\infty e^{10\text{kcal}/RT}} = \frac{1}{KAPG} + \frac{[A]}{APG} \qquad (2.11b)$$

The experimental results of Schnecko *et al.*[17] are shown in Fig. 2.22

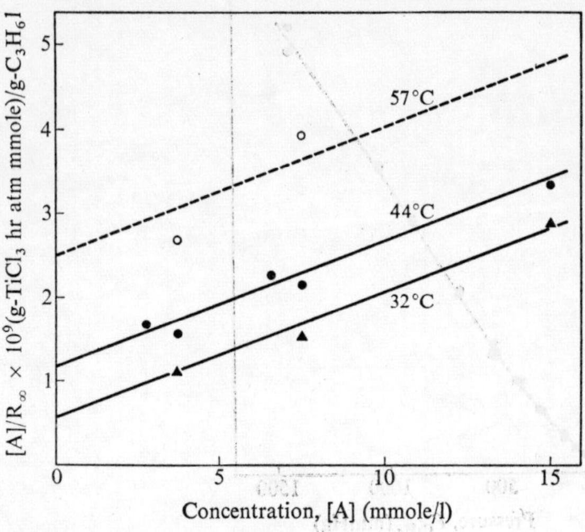

Fig. 2.21. Langmuir plot of stationary rate against $AlEt_3$ concentration.[9] ($[TiCl_3]$ = 2 g/l, 250 ml *n*-heptane)

and 2.23. Although both results were broadly in agreement, the values of K that were observed in each case did not coincide.

We may note here the dependence of decay type kinetic curves on $AlEt_3$ concentration. As discussed in the previous section, k_3 is effectively independent of $AlEt_3$ concentration, and consequently the left-hand side of Eq. 2.7*a* should also be independent of concentration, so that $\bar{R}_0$ and R should then have the same concentration dependence as R_∞. This has been confirmed in the case of $\bar{R}_0$ by the work of Keii and Soga.[9] It is therefore valid to use integral rates in determining the concentration dependence of the polymerization rate, R, as in the work of Schnecko *et al.*[17]

Ingberman *et al.*[15] studied the influence of $AlEt_2Cl$ concentration on polymer yield (R') after propylene polymerization for 2.5 hr with $TiCl_3$, using a reactor of the type shown in Fig. 1.4. They proposed the following rate law

$$R' = Ae^{-12\text{kcal}/RT}G[\text{M}]\left(\frac{K'[\text{A}']^{1/2}}{1 + K'[\text{A}']^{1/2}}\right) \tag{2.12a}$$

where [M] is the monomer concentration, and [A′] the corrected $AlEt_2Cl$

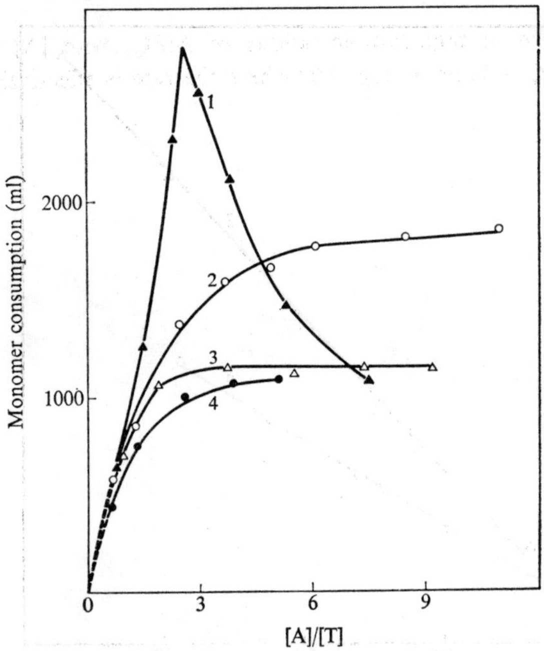

Fig. 2.22. Monomer consumption (C_E, C_{Pr}) *vs* [A]/[Ti] for various catalyst systems according to Schnecko *et al.*[17] (1, $[TiCl_4] = 1.3 \times 10^{-3}$ mole/l with AlR_3, $P_E = 350$mmHg, 50°C; 2, $[TiCl_3] = 2.6 \times 10^{-3}$ mole/l with AlR_3, $P_E = 450$ mmHg, 50°C; 3, $[TiCl_3] = 4.0 \times 10^{-3}$ mole/l with AlR_3, $P_{Pr} = 355$ mm Hg, 56°C; 4, $[TiCl_4] = 5.1 \times 10^{-3}$ mole/l with $AlRCl_2$, $P_E = 450$ mmHg, 50°C.)

concentration. They used the correction factor given in Eq. 2.12*b*

$$[A'] = [A] - \alpha \qquad (2.12b)$$

where α was constant for each run. It was supposed that α represented the amount of $AlEt_2Cl$ destroyed by impurities in the polymerization system, and its values were so determined that the functional form of Eq. 2.12*a* could account for the experimental results. The data presented by them are quoted in Table 2.2, for which $\alpha = 0.08$ mmole. However, the original data can be explained in terms of a Langmuir dependence on [A], without any correction. The linear plots of $1/[A']^{1/2}$ against $1/R'$ (open circles) obtained by Ingberman *et al.* are shown in Fig. 2.24 and 2.25, and those of $1/[A]$ against $1/R'$ are indicated by closed circles.

As can be seen from the functional forms of the two equations

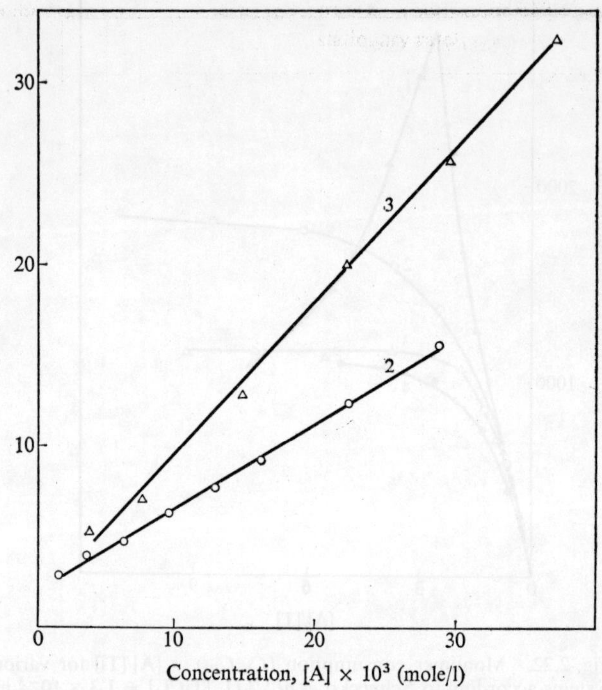

Fig. 2.23. Langmuir plot of rate against AlR_3 concentration.[17] Conditions as for Fig. 2.22.

TABLE 2.2 Effect of diethylaluminium chloride concentration on propylene polymerization rate†

$TiCl_3$ (mmole)	$AlEt_2Cl$ (mmole)	[A]/[Ti]	Polymer yield (g/mmole-Ti)	Ether solubles (%)
2.10	0.153	0.070	0.22	
2.08	0.384	0.160	3.61	1.5
2.06	0.538	0.262	4.26	2.0
2.04	0.940	0.461	4.75	0
2.12	1.09	0.515	4.97	0
1.85	1.90	1.03	5.70	1.8
1.98	11.0	5.52	6.70	1.0
2.12	24.5	11.6	6.90	0.7

† 40°C, 7.5 psig propylene, 2.5 hr, 1.5 l *n*-heptane.

(Eq. 2.13*a* and *b*), they may be clearly distinguished at low concentrations

$$R = R^\circ\left(\frac{K[A]}{1 + K[A]}\right) \tag{2.13a}$$

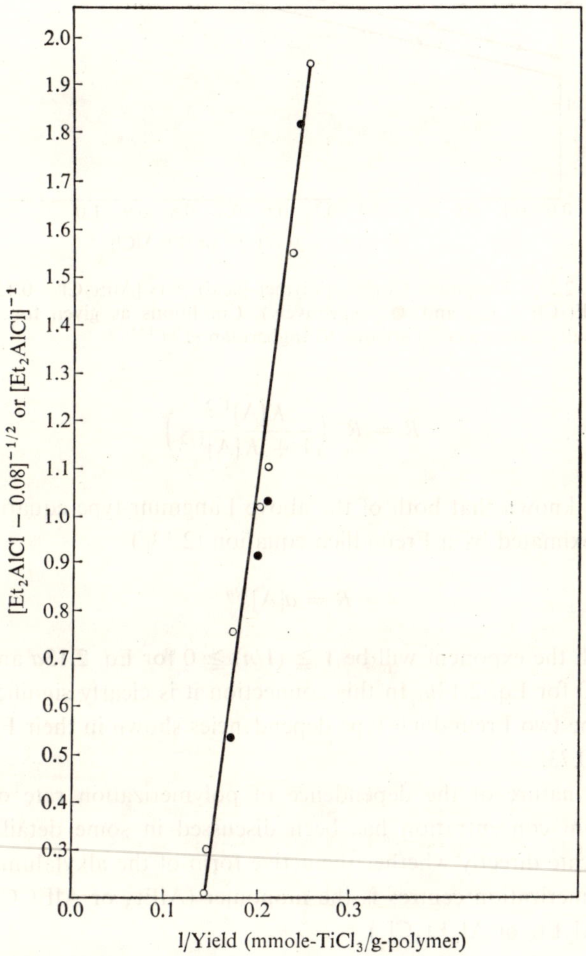

Fig. 2.24. Langmuir plots of (polymer yield)$^{-1}$ *vs* [$AlEt_2Cl$ −0.08]$^{-1/2}$ and (polymer yield)$^{-1}$ *vs* [$AlEt_2Cl$]$^{-1}$ (○ and ● respectively). [$AlEt_2Cl$] refers to the total amount in 1.5 l of *n*-heptane solvent. (P_{Pr} = 7.5 psig at 47°C. Polymerization time, 2.5 hr). The results shown as (○) are due to Ingberman *et al.*[15]

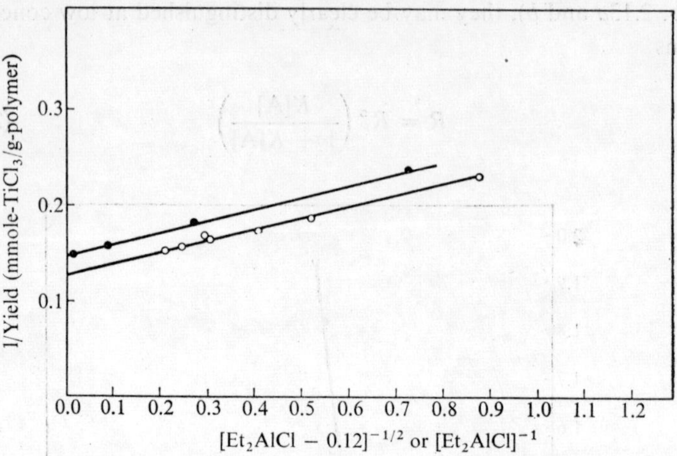

Fig. 2.25. Langmuir plot of (polymer yield)$^{-1}$ *vs* $[AlEt_2Cl - 0.12]^{-1/2}$ and $[AlEt_2Cl]^{-1}$ (○ and ● respectively). Conditions as given for Fig. 2.24. Results shown as (○) are due to Ingberman *et al.*[15]

$$R = R^\circ \left(\frac{K[\mathrm{A}]^{1/2}}{1 + K[\mathrm{A}]^{1/2}} \right) \tag{2.13b}$$

It is well known that both of the above Langmuir-type equations may be approximated by a Freundlich equation (2.13*c*)

$$R = a[\mathrm{A}]^{1/n} \tag{2.13c}$$

for which the exponent will be $1 \geqq (1/n) \geqq 0$ for Eq. 2.13*a* and $1/2 \geqq (1/n) \geqq 0$ for Eq. 2.13*b*. In this connection it is clearly significant that one of the two Freundlich type dependencies shown in their Fig. 1 has $(1/n) = 0.73$.

The nature of the dependence of polymerization rate on alkylaluminium concentration has been discussed in some detail since it can indicate directly whether the active form of the alkylaluminium at the polymerization centres is the monomer ($AlEt_3$ or $AlEt_2Cl$) or the dimer (Al_2Et_6 or $Al_2Et_4Cl_2$).

2.4.3 Langmuir-Hinshelwood rate laws

The rate laws given in Eq. 2.10, 2.11 and 2.12 are all of the Rideal type, as first proposed for heterogeneous catalysis by Twigg and

Rideal,[18] and elaborated by Eley and Rideal.[19] However, the Langmuir-Hinshelwood equations, originating in the work of Langmuir on adsorption and heterogeneous catalysis,[20] are also commonly applicable in this field. Both types of rate law are valuable in that they correspond to simple physical processes connected with the mechanism of surface reactions.

Langmuir-Hinshelwood rate laws have been proposed by Vessely *et al.*[21] for propylene polymerization with $TiCl_3/AlEt_3$, by Schindler[22] for ethylene polymerization using $TiCl_4/AlEt_2Cl$, and by Tait and McKenzie[23] for 4-methylpentene-1 with $VCl_3/Al(i\text{-}Bu)_3$. Vessely's rate law for propylene polymerization with $TiCl_3$ was

$$R_\infty = \frac{kK_MK_A[M][A]}{(1 + K_M[M] + K_A[A])^2} \tag{2.14a}$$

where R_∞ is the stationary rate per unit mass of $TiCl_3$ and [M] represents the monomer concentration. The dependence of the rate, R_∞, on the concentration of $AlEt_3$ should have a maximum in this type of equation, and the concentration $[A]_{max}$ at the maximum rate can be obtained by differentiation of Eq. 2.14*a* with respect to [A].

$$K_A[A]_{max} = 1 + K_M[M] \tag{2.14b}$$

Hence we can obtain the following expression for the maximum rate of polymerization:

$$(R_\infty)_{max} = \frac{kK_M[M]}{4(1 + K_M[M])} \tag{2.14c}$$

The value of K_M, calculated from the experimental results shown in Fig. 2.26, was 0.163 l/mole. Using this value in Eq. 2.14*b*, K_A was found to be 21.2 l/mole. A similar rate law can be obtained from the experimental results of Tait *et al.*[23] shown in Fig. 2.27, though the authors proposed the rate law given in Eq. 2.15, which was applicable only for high concentrations of $Al(i\text{-}Bu)_3$.

$$R = \frac{kK_M[M]}{1 + K_M[M] + K_A[A]} \tag{2.15}$$

From the results shown in Fig. 2.27, it can be found that $K_A = 40 \sim 60$ l/mole at 40°C. Experimental values of K_A and K_M reported by several authors are summarized in Table 2.3.

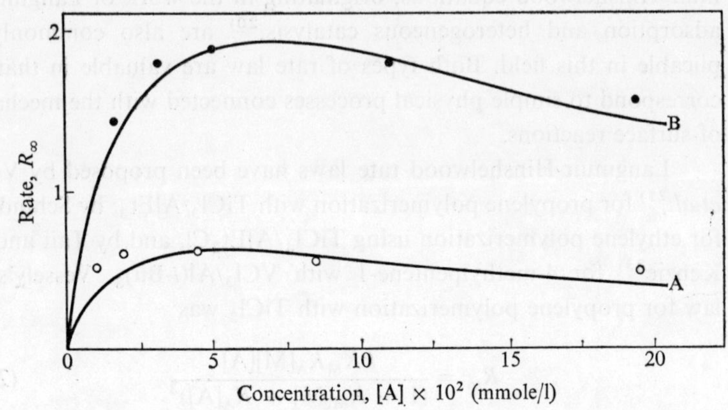

Fig. 2.26. Theoretical and experimental dependence of polymerization rate on $AlEt_3$ concentration. (A, [M] = 0.58 mole/l at 50° C; B [M] = 2.3 mole/l at 50° C). Data due to Vessely.[21)]

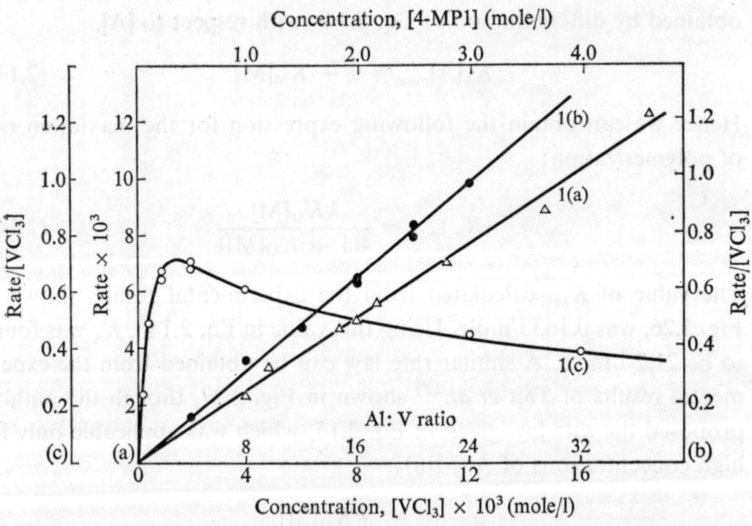

Fig. 2.27. Variation of polymerization rate with [A]/[V] according to Tait *et al.*[23)] ($[VCl_3] = 8.8 \times 10^{-3}$ mole/l, [4-MP-1] = 2.0 mole/l, (○); [A]/[V] = 2, [4-MP-1] = 2.0 mole/l, (▲); $[VCl_3] = 8.8 \times 10^{-3}$ mole/l, [A]/[V] = 2, (●).) [MP = methyl pentene].

TABLE 2.3 Experimental values of K_A and K_M reported by various authors

Authors	T (°C)	K_A (l/mole)	E_A (kcal/mole)	K_M (l/mole)	Type of rate law
Keii and Soga[9]	32	280			
	44	170	12.0		Rideal type
	57	60			
Schnecko *et al.*[17]	56	600			
Vessely[21]	50	21.2	35.0	0.163	Langmuir-Hinshelwood type
Tait *et al.*[23]	40	40~60		6~9	

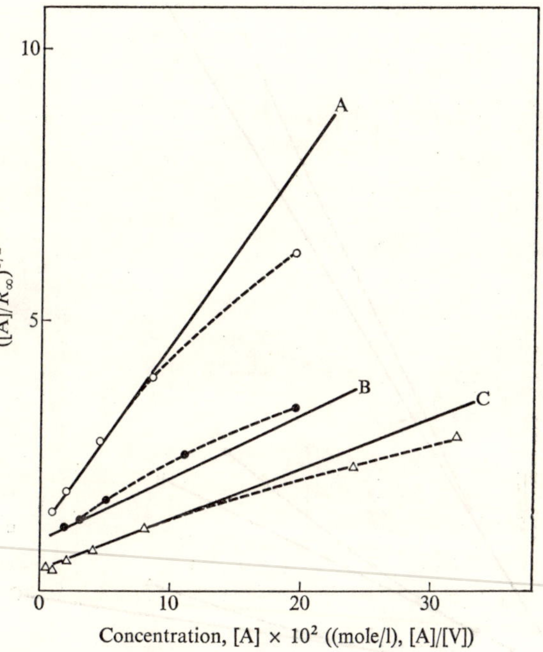

Fig. 2.28. Applicability of a Langmuir-Hinshelwood rate law to the data of Vessely[21] and Tait *et al.*[23] (see Eq. 2.15*a*. A, B from Fig. 2.26, and C from Fig. 2.27.)

The applicability of Langmuir-Hinshelwood rate laws may be tested in any particular case by seeking a linear relationship between $([A]/R_\infty)^{1/2}$ and [A], as can be seen from the linear form of Eq. 2.14*a*.

$$\left(\frac{[A]}{R_\infty}\right)^{1/2} = \frac{1 + K_M[M]}{(kK_AK_M[M])^{1/2}} + \left(\frac{K_A}{kK_M[M]}\right)^{1/2}[A] \qquad (2.15a)$$

Fig. 2.28 illustrates the application of Eq. 2.15*a* to the experimental data of Vessely *et al.*[21] (A and B) and Tait *et al.*[23] (C). Fig. 2.29 shows the corresponding relationship between $[A]/R_\infty$ and [A], derived from the Rideal type rate law Eq. 2.11*a*, given below in linear form.

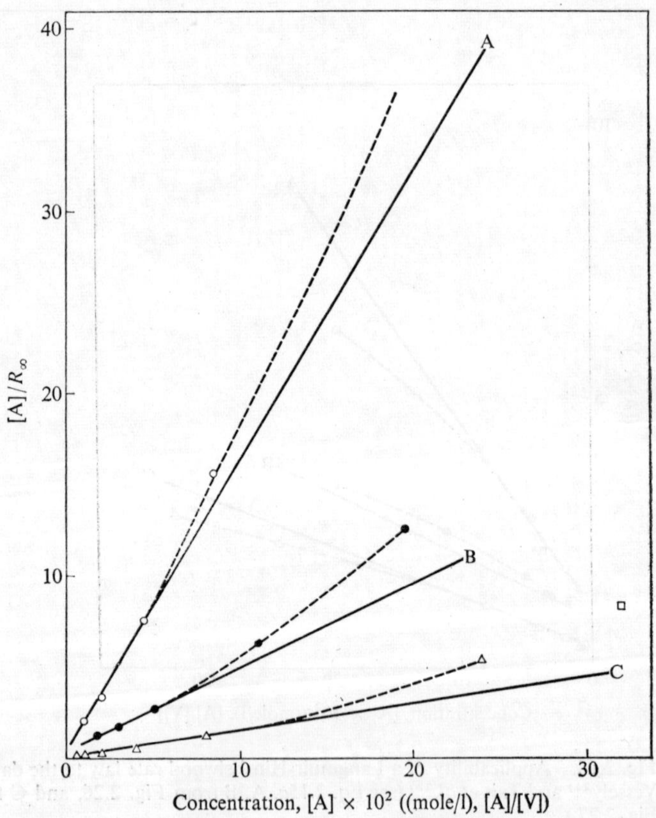

Fig. 2.29. Applicability of a Rideal rate law (Eq. 2.15 *b*) to the data of Vessely[21] and Tait *et al.*[23]

$$\frac{[\mathrm{A}]}{R_\infty} = \frac{1}{KAP} + \frac{[\mathrm{A}]}{AP} \tag{2.15b}$$

It can be seen that Langmuir-Hinshelwood and Rideal-type rate laws are indistinguishable at low $AlEt_3$ concentrations and given monomer concentrations.

2.4.4 Relationship between Langmuir-Hinshelwood and Rideal rate laws

In general, Rideal-type rate laws have been found by workers using low $AlEt_3$ concentrations and low propylene pressures. If the Langmuir-Hinshelwood rate law proposed by Vessely *et al.* is valid under these conditions, it may be applied to the data of Keii and Soga,[9] where the highest concentrations used were $[\mathrm{A}] = 40 \times 10^{-3}$ mole/l, i.e.

$$K_\mathrm{M}[\mathrm{M}] \ll K_\mathrm{A}[\mathrm{A}] < 21.2 \times 40 \times 10^{-3} < 1$$

In this case we may approximate Eq. 2.14*a* as follows

$$\begin{aligned} R_\infty &= k\frac{K_\mathrm{M}K_\mathrm{A}[\mathrm{M}][\mathrm{A}]}{(1 + K_\mathrm{M}[\mathrm{M}] + K_\mathrm{A}[\mathrm{A}])^2} \\ &\simeq k\frac{K_\mathrm{M}K_\mathrm{A}[\mathrm{M}][\mathrm{A}]}{(1 + K_\mathrm{A}[\mathrm{A}])^2} \\ &\simeq \frac{k}{2}\frac{K_\mathrm{M}[\mathrm{M}](2K_\mathrm{A}[\mathrm{A}])}{(1 + 2K_\mathrm{A}[\mathrm{A}])} \end{aligned} \tag{2.16a}$$

Monomer concentration [M] is given by Henry's law

$$[\mathrm{M}] = HP \tag{2.16b}$$

and we therefore obtain, putting $2K_\mathrm{A} = K$,

$$R_\infty = \left(\frac{kHK_\mathrm{M}}{2}\right)P\left(\frac{K[\mathrm{A}]}{1 + K[\mathrm{A}]}\right) \tag{2.16c}$$

which is of the same form as Rideal-type equations. The values of K obtained by Keii and Soga[9] ($K = 60$ l/mole at 57 °C) give some support to the relation $2K_A = K$ when compared with Vessely's experimental value for K_A of 21.2 l/mole at 50 °C. However, the present author is reluctant to recognize his Rideal-type rate equation as a valid approximation of Vessely's Langmuir-Hinshelwood equation. The latter equation, obtained by Vessely from experimental data on rate maxima, does not appear to coincide with the experimental results obtained at lower concentrations of $AlEt_3$, as can be seen from Fig. 2.26, and furthermore, the approximation of the Langmuir-Hinshelwood equation in this way (Eq. 2.16*a*) would be expected to give rise to an extraordinarily small frequency factor for the polymerization, due to the assumption that $K_M[M] \ll 1$.

2.5 INFLUENCE OF $TiCl_3$ ON POLYMERIZATION KINETICS

2.5.1 The nature of $TiCl_3$ catalysts

There are several kinds of $TiCl_3$ catalysts commercially available, all of which are prepared by the reduction of $TiCl_4$ with H_2, Al, Ti or $AlEt_3$, followed by activation by grinding or milling. According to Natta *et al.*[24] there are four crystalline modifications of $TiCl_3$, known as α, β, γ and δ forms, of which the α form is the best known. It may be prepared by the reduction of $TiCl_4$ with H_2 (above 500 °C), with Al (above 250 °C) or with metals (above 400 °C).[25] Reduction at lower temperatures yields the β form.[26] The γ form is obtained by heating the β form above 150 °C, or by reducing $TiCl_4$ with organoaluminium compounds at 150–200 °C,[27] and the δ form either by prolonged grinding of α or γ forms, or by grinding the α form with $AlCl_3$.[24]

The crystal structures of the various modifications are illustrated in Fig. 2.30. $TiCl_3$ is a typical ionic crystal, having a high melting point (it decomposes to $TiCl_2$ and $TiCl_4$ at 450 °C, and sublimes at 830 °C in $TiCl_4$ vapour) and a high specific gravity of 2.69, as might be expected from its lack of porosity and low specific surface area. It is soluble in polar solvents such as alcohols or tetrahydrofuran, but insoluble in hydrocarbons.

Most commercial $TiCl_3$ occludes a small amount of $TiCl_4$, which is readily removed by evacuation at 200 °C. The specific surface areas of commercial active (ground) catalysts are in the range 10–40 m^2/g, while those of unground samples, e.g. A-, H-grades of Stauffer Chemi-

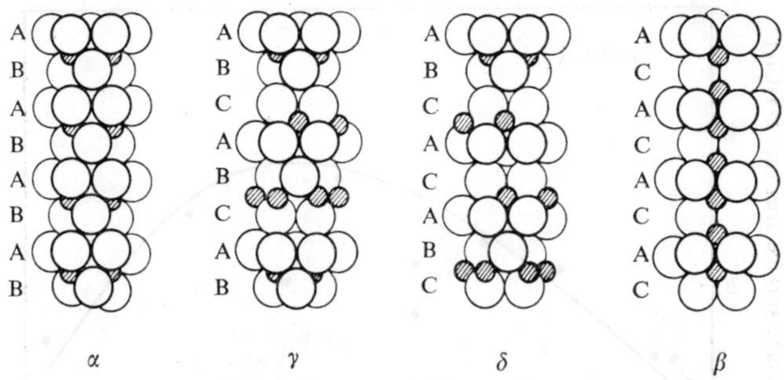

Fig. 2.30. Crystal structures of the four modifications of $TiCl_3$.

cal Co., are lower than 10 m^2/g. The highest specific surface area reported for α-$TiCl_3$ is about 100 m^2/g, which may be approaching the upper limit for crystalline materials of this type.

2.5.2 Effect of grinding on $TiCl_3$ surface area

As a general practice, $TiCl_3$ catalysts are ground before use in a ball or jet mill, or similar device, in order to increase their activity for polymerization. It is known that grinding increases the surface area, as well as changing the crystal structure into the δ form, but the change in the surface area is not a simple one.[27)]

Two samples of α-$TiCl_3$ (H-grade, Stauffer Chemical Co., prepared by reduction with H_2, with an initial surface area of 1.4 m^2/g, and TA-grade of Tohō Titanium Co., prepared by reduction with Al and with initial surface area 4.6 m^2/g) were ground at 100 rpm under a nitrogen atmosphere in a porcelain ball mill of 9 cm diameter with ten 1.8 cm diameter balls and five 2.3 cm diameter balls. An initial sample size of 10–40 g was used, and 1g samples of $TiCl_3$ were withdrawn every 20 hr to determine the surface area and polymerization activity. The surface area was found by the BET method of nitrogen adsorption at liquid nitrogen temperature.[28)] No chemisorption of nitrogen on $TiCl_3$ is known to occur at this temperature. The observed dependence of specific surface area on the grinding time is shown in Fig. 2.31 and 2.32.

It is well known that the grinding effect of such devices is dependent on the amount of powder in the mill, as well as the duration of grinding, and it is therefore preferable to convert the ordinate in Fig. 2.31 and 2.32 into grinding duration per unit weight of $TiCl_3$, i.e. into "grinding

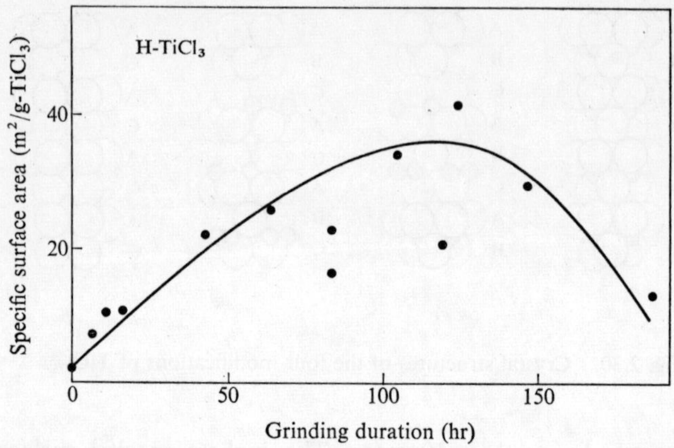

Fig. 2.31. Dependence of specific surface area on grinding duration for H-$TiCl_3$, according to Keii *et al.*[27]

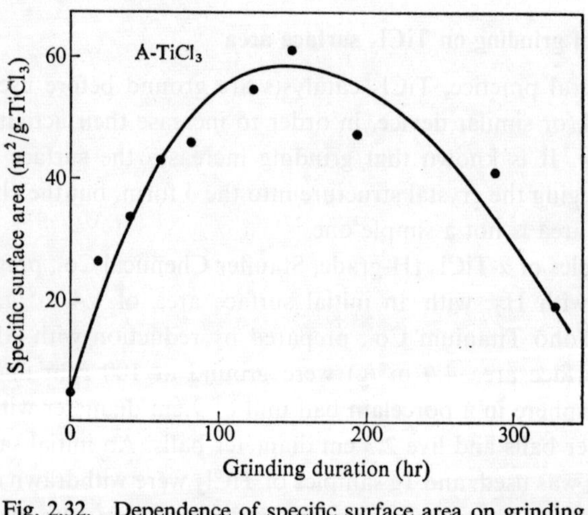

Fig. 2.32. Dependence of specific surface area on grinding duration for A-$TiCl_3$.[27]

work", E_t, as defined below.

$$E_t = \varepsilon \int_0^t \frac{dt}{G(t)} \tag{2.17}$$

$G(t)$ is the mass of $TiCl_3$ at grinding time t, and ε is the effective work per unit mass and time. The latter is a constant for a particular sample of $TiCl_3$. Fig. 2.33 and 2.34 show the experimental dependence of the specific surface area, *S*, on the grinding work E_t. The initial stages of grinding can be represented by Rittinger's law, $E_t = \text{const.} \times (S_T - S_0)$ for H-$TiCl_3$ and by Tanaka's law,[29] $S_T = S_\infty(1 - e^{-kEt})$ for A-$TiCl_3$.

The wave-like change shown in Fig. 2.34 has also been found to

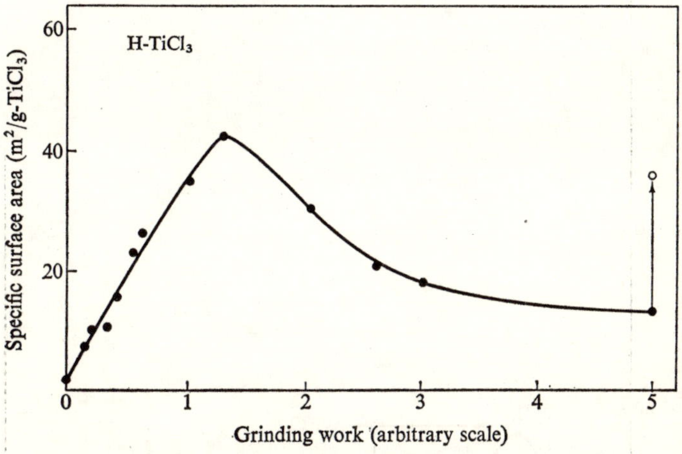

Fig. 2.33. Relationship between specific surface area and grinding work for H-$TiCl_3$.[27]

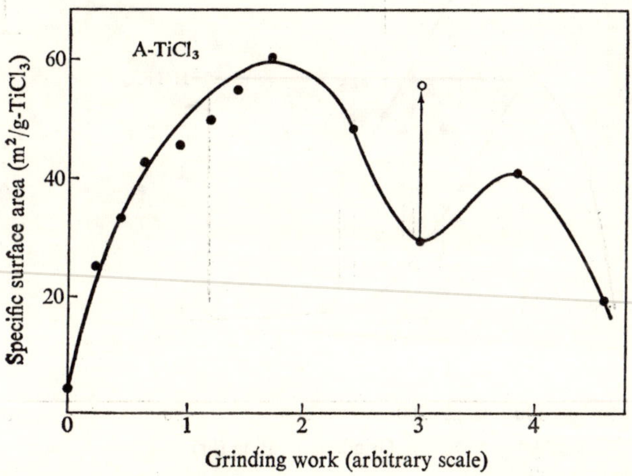

Fig. 2.34. Relationship between specific surface area and grinding work for A-$TiCl_3$.[27]

occur during the grinding of kaolinite,[30)] sodium chloride,[31)] iron powder[32)] etc., and a similar result has been reported with a sample of α-$TiCl_3$[33)] ($TiCl_3/0.2AlCl_3$, prepared by reducing $TiCl_4$ with a mixture of Ti and Al). It has been shown by Keii *et al.* that the appearance of a maximum in the surface area is due, at least in their case,

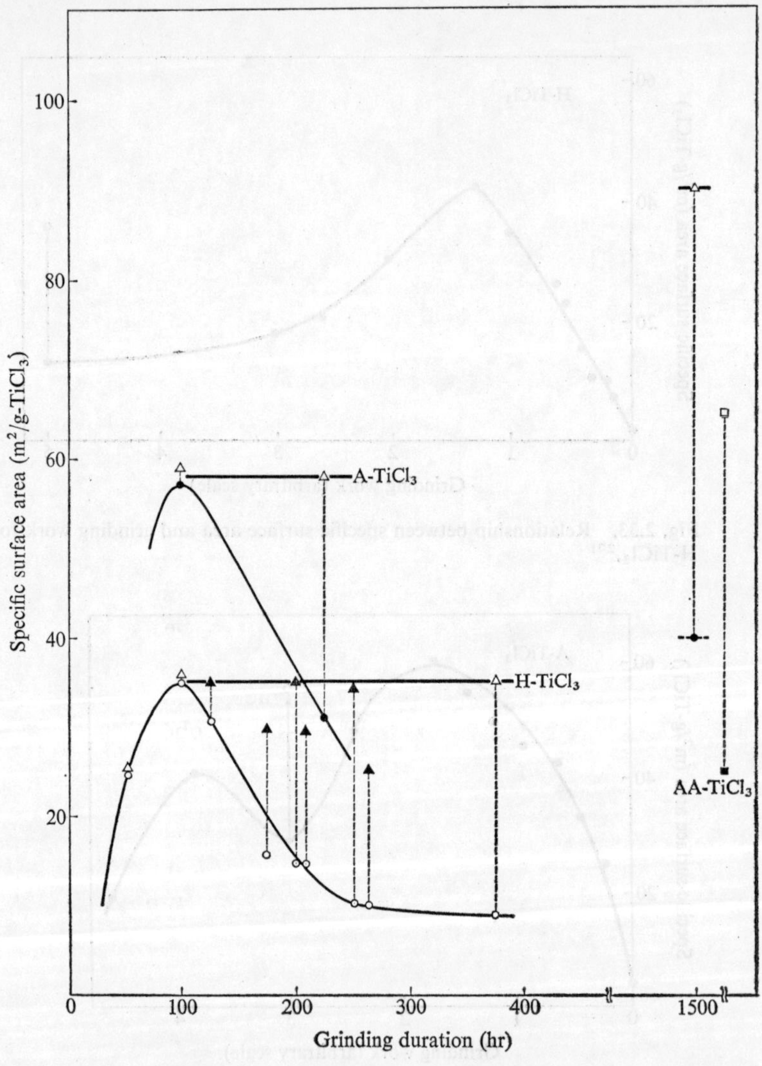

Fig. 2.35. Changes in specific surface area on grinding, showing the effect of solvent treatment.[27)]

mainly to the agglomeration of $TiCl_3$ particles. On treating ground samples of $TiCl_3$ with a small amount of toluene or *n*-heptane, recovery of surface area was observed in both samples of $TiCl_3$, as indicated in Fig. 2.33 and 2.34. A systematic study of surface area recovery of ground samples has been carried out by Keii, Suzuki and Soga.[27] The ground sample was agitated under nitrogen with 10 ml of toluene or *n*-heptane for 30 min at room temperature. The liquid phase was then removed, initially with a syringe, and finally by complete evacuation, and the surface area of the sample was measured again. Weight loss during this treatment did not exceed 5%. The results are shown in Fig. 2.35.

As can be seen from Fig. 2.35, the solvent recovery effect was particularly marked in the case of AA-grade $TiCl_3$. The nature of the solvent did not appear to affect the recovery, indicating that the effect is due to de-agglomeration of $TiCl_3$ particles, and not to the dissolution of $AlCl_3$ from the $TiCl_3$.

2.5.3 Influence of $TiCl_3$ grinding on polymerization kinetics

One of the effects of $TiCl_3$ grinding on polymerization kinetics was first pointed out by Natta *et al.*[5,6] who compared the kinetic curves obtained with ground and unground samples of α-$TiCl_3$ as mentioned in (2.1.1). The observed kinetic curves are shown in Fig. 2.36. Unfortunately the specific surface areas of the samples were not indicated, although approximate particle sizes were given. On the basis of these results, Natta *et al.* proposed their theory of particle size adjustment due to the mechanical action of polymer chains. Keii and Akiyama[34] examined the effect of grinding α-$TiCl_3$ (H-grade, surface area 3.5 m^2/g) on propylene polymerization. The kinetic curves obtained are shown in Fig. 2.37, in which the polymerization rates are normalized with m^2. It can be seen that the stationary rate (m^{-2}) remains unchanged, whereas the initial rate (m^{-2}) increases with catalyst grinding duration. In this case the same amount of $TiCl_3$ was used in each case so that the duration could be taken as the grinding work. The two kinetic curves obtained using old samples (stored for about one month under nitrogen, after grinding) show a marked decrease in the initial rate of polymerization, though both the surface area and the stationary rate were unchanged. All the kinetic curves could be represented by Eq. 2.7*a* or *b*.

More quantitative studies have been carried out by Keii, Suzuki

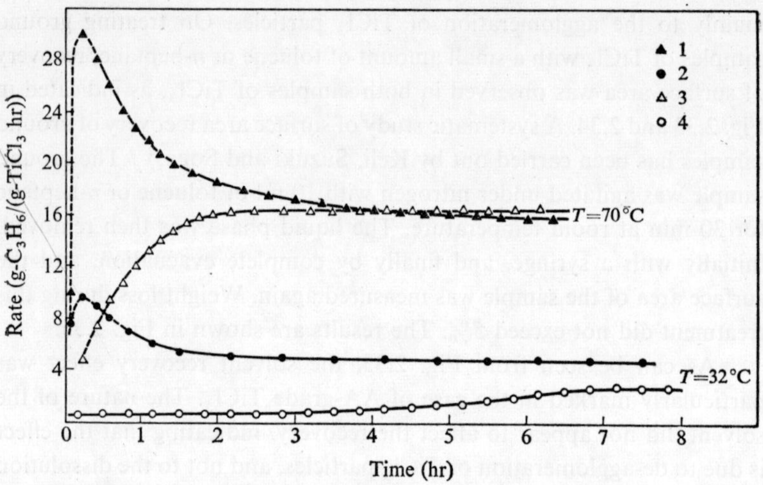

Fig. 2.36. Effect of grinding on the shapes of the kinetic curves.[4)] 1 and 2 show ground $TiCl_3$ (sample A), particle size $\leqslant 2\,\mu$; 3 and 4 show unground $TiCl_3$ (sample A), particle size 1–10 μ. $P = 1450$ mmHg.

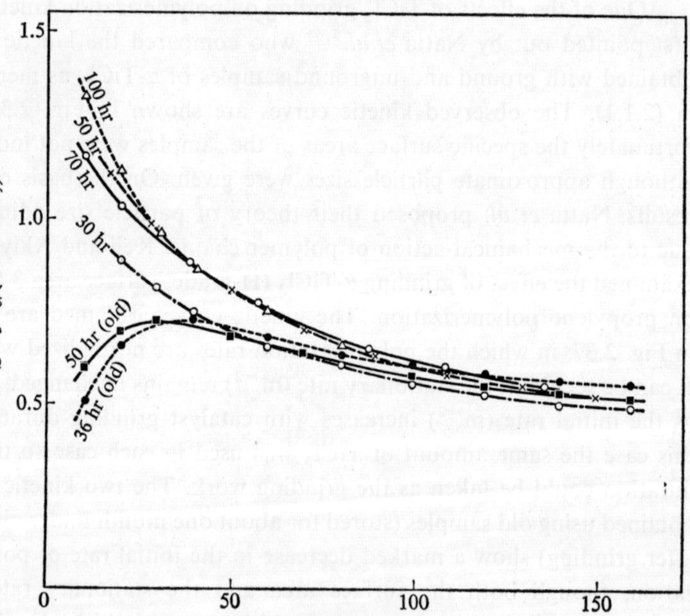

Fig. 2.37. Kinetic curves obtained by Keii and Akiyama,[34)] using ground H-grade $TiCl_3$.

and Soga.[27] Kinetic curves obtained with ground H-$TiCl_3$ and $AlEt_3$ are shown in Fig. 2.38, where the polymerization rate is expressed in g-C_3H_6/(g-$TiCl_3$ hr atm). It can be seen from Fig. 2.37 and 2.38 that the initial rate increases with the grinding work until the surface area of the catalyst reaches its maximum value, and thereafter the initial rate decreases with further grinding. The stationary rate is also proportional to the surface area of the catalyst, until this reaches its maximum, as shown in Fig. 2.39, and even after this, the proportionality is maintained if the "true" surface area of the catalyst is considered, i.e. the value found after de-agglomeration by treatment with solvent. Similar results have been obtained with A-$TiCl_3$, the grinding behavior of

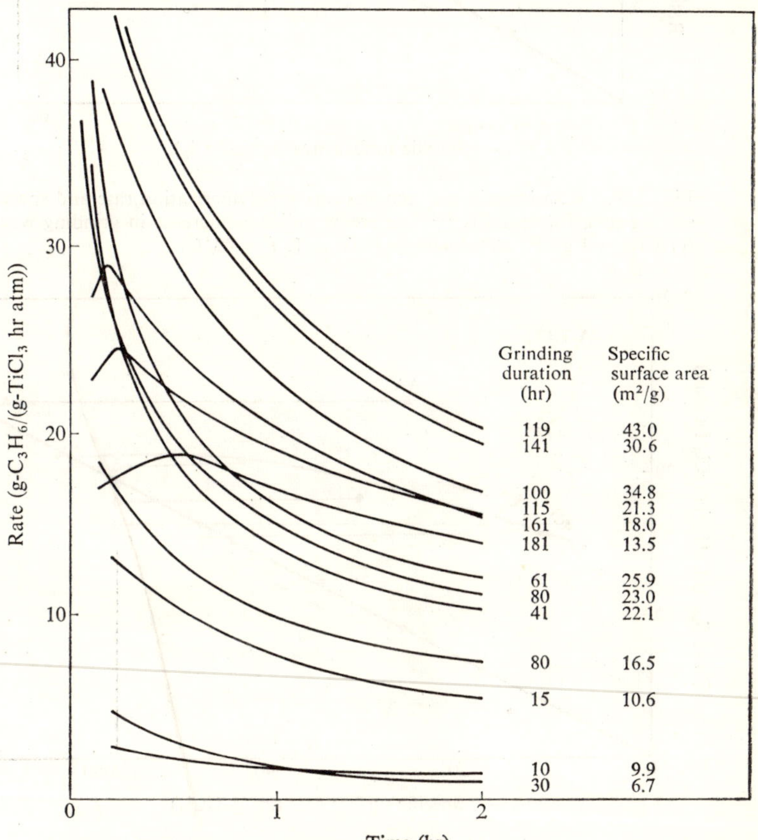

Fig. 2.38. Kinetic curves observed with various samples of ground H-$TiCl_3$.[27] ([$TiCl_3$] = 1 g/250 ml of toluene, [A]/[Ti] = 1, T = 43°C).

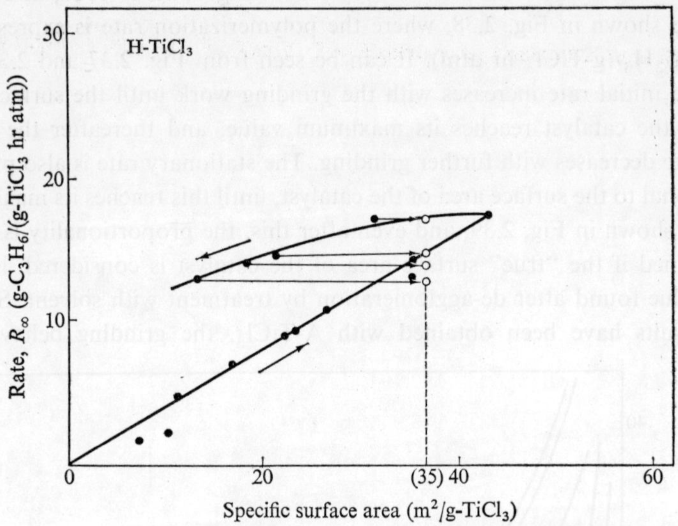

Fig. 2.39. Relationship between stationary polymerization rate and specific surface area for H-$TiCl_3$.[27] The arrow indicates increase in grinding work. ([$TiCl_3$] = 1 g/250 ml toluene, [A]/[Ti] = 1, T = 43°C)

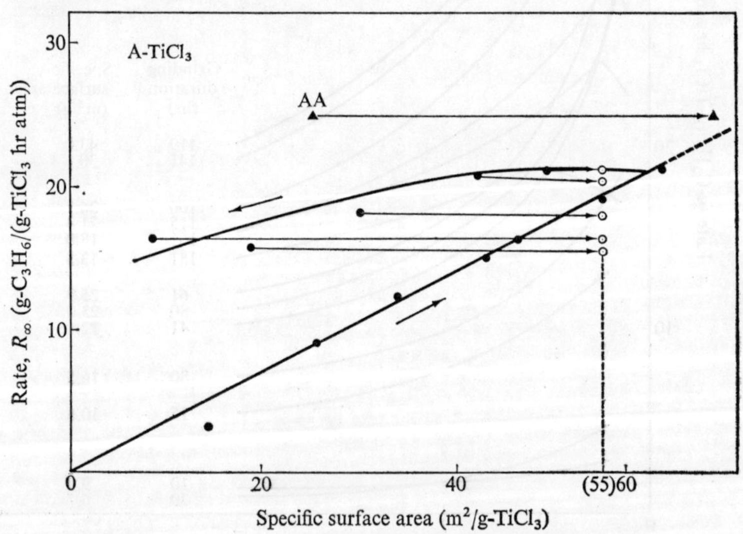

Fig. 2.40. Relationship between stationary polymerization rate and specific surface area for A-$TiCl_3$.[27] Conditions are as in Fig. 2.39.

which was shown in Fig. 2.32 and 2.34. These results are summarized, together with those found for AA-grade commercial catalyst (Stauffer Chemical Co.) in Fig. 2.40. It can be seen that the slopes of the lines in Fig. 2.39 and 2.40 are very similar. Comparable results were found during the polymerization of propylene with ground $TiCl_3$ and $AlEt_2Cl$, as shown in Fig. 2.41 and 2.42.

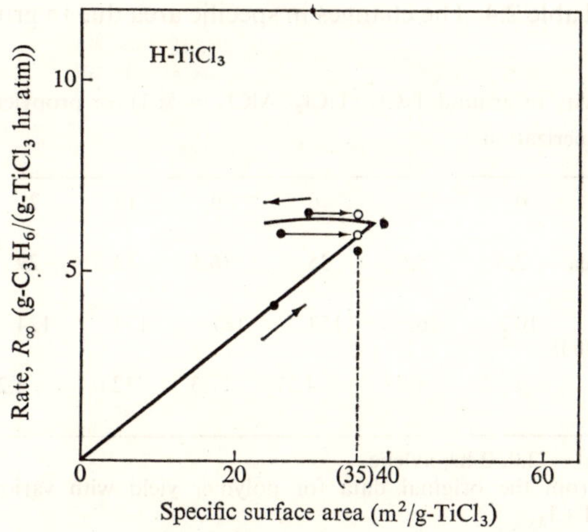

Fig. 2.41. Relationship between stationary polymerization rate and specific surface area for $AlEt_2Cl$/H-$TiCl_3$.[27] Conditions are as in Fig. 2.39.

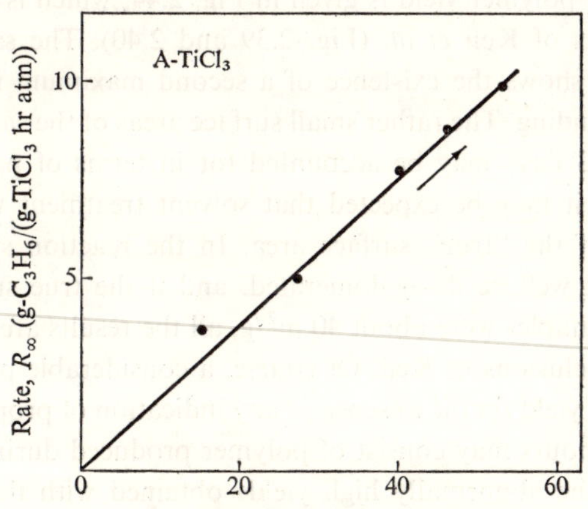

Fig. 2.42. Relationship between stationary polymerization rate and specific surface area for $AlEt_2Cl$/A-$TiCl_3$.[27] Conditions are as in Fig. 2.39.

Because of these changes in the rate of polymerization with time and grinding work, it is apparent once more that the use of integral rates is unlikely to prove meaningful. The data of Tornqvist[33)] have sometimes been quoted as evidence for the independence of catalyst activity and surface area. Samples of $TiCl_3$ ($TiCl_3/0.2AlCl_3$) were ground with flint pebbles in a porcelain vessel, and were then examined for changes in surface area and polymerization activity. Some of these data are shown in Table 2.4. The changes in specific area due to grinding

TABLE 2.4 Activity of ground $TiCl_3$ ($TiCl_3 : AlCl_3 = 5 : 1$) for propylene polymerization†1

Grinding duration (day)	0	2	6	9	13	21
Specific surface area (m^2/g)	2.5	5.6	35	16.5	14.5	29
Polymer yield†2 ($g\text{-}C_3H_6/(g\text{-}TiCl_3\ hr)$)	10.3	26.5	153	187	174	124
Polymer yields†2 ($g\text{-}C_3H_6/(hr\ m^2)$)	4.12	4.73	4.37	11.3	12.0	4.28

†1 75°C, [A]/[Ti] = 2.0, 2 hr, xylene.
†2 Calculated from the original data for polymer yield with various amounts of $TiCl_3$.

(shown in Fig. 2.43) are similar to those found by Keii *et al.*,[27)] and the effect of grinding on polymer yield is given in Fig. 2.44, which is again similar to the results of Keii *et al.* (Fig. 2.39 and 2.40). The sample ground for 21 days shows the existence of a second maximum in the surface area after grinding. The rather small surface areas of the samples ground for 9 and 13 days may be accounted for in terms of particle agglomeration, and it may be expected that solvent treatment would result in recovery of the "true" surface area. In the reaction system such powders might well be de-agglomerated, and if the true surface areas of these two samples were about 40 m^2/g, all the results are compatible with the conclusions of Keii. Of course, a considerable part of Tornqvist's polymer yield (in the absence of any indication of propylene pressure) after two hours may consist of polymer produced during the decay period, and the abnormally high yields obtained with the two samples referred to may be partly due to the increase in the initial rate with grinding.

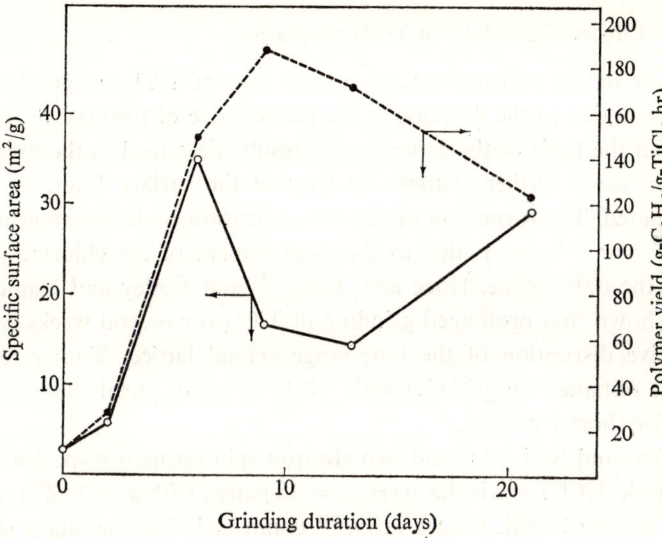

Fig. 2.43. Changes in specific surface area and polymerization activity of $TiCl_3 \cdot 0.2\ AlCl_3$ on grinding.[33)]

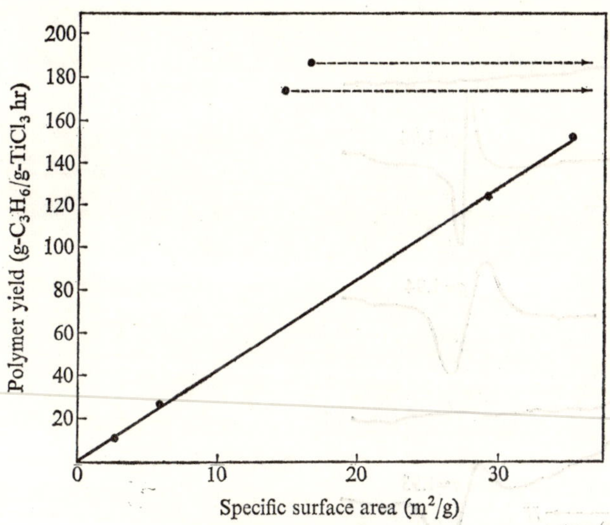

Fig. 2.44. Relationship between polymerization yield and specific surface area, calculated from the data in Fig. 2.43.

2.5.4 Effects of grinding on $TiCl_3$ properties

The increase in the specific surface area of $TiCl_3$ on grinding is, of course, due to the decrease in the particle size of the crystals. However, on the basis of the experimental results described in the previous section, some further changes, at least in the surface layer, may be anticipated. The formation of the δ modification on intensive grinding of the α or β forms is due to the displacement of the chlorine plane along the (001) plane. Hoeg and Ashby[35)] and Ketley and Daniels[36)] have shown that prolonged grinding of $TiCl_3$ for several weeks causes extensive disruption of the long-range crystal lattice. Similar results can be obtained by grinding with $AlCl_3$ or alkali metal salts such as LiCl for shorter periods.

Ono and Keii [37)] found two electron spin resonance spectra from AA-grade $TiCl_3$; one is sharp (H_{mse} = 15 gauss) with g = 1.94, and the other is very broad. Okura *et al.*[38)] confirmed that the sharp signal appeared after grinding. The spectra measured before and after grinding of a 15 g sample for 100 hr are shown in Fig. 2.45. It was confirmed that the signal is an effect of grinding and not due to aluminium oxide

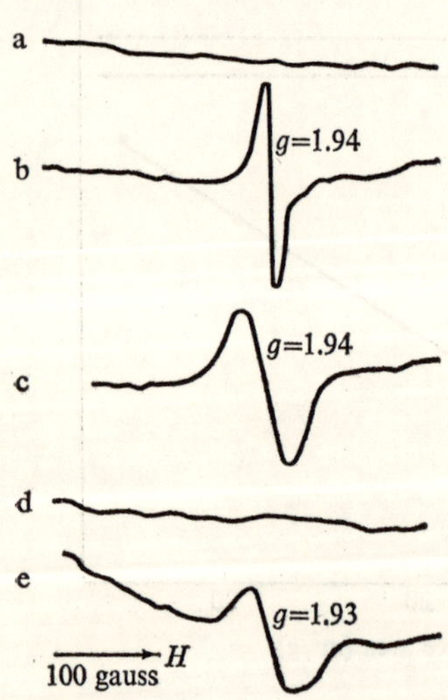

Fig. 2.45. ESR spectra of $TiCl_3$ after grinding. (a, A-$TiCl_3$; b, ground A-$TiCl_3$; c, AA-$TiCl_3$; d, H-$TiCl_3$; e, ground H-$TiCl_3$).[38)]

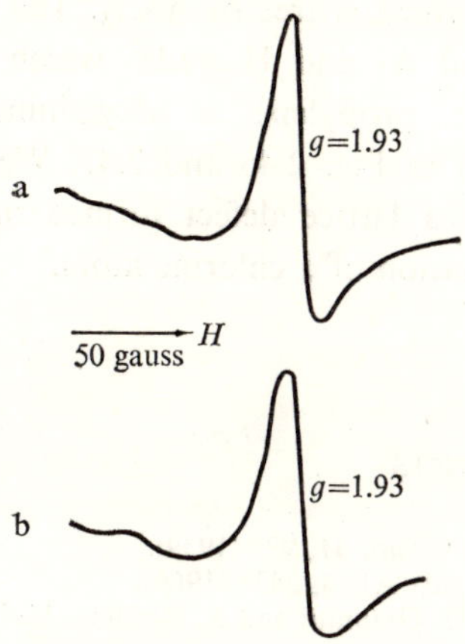

Fig. 2.46. ESR spectra of A-$TiCl_3$ after addition of olefin (a, ethylene; b, propylene).[38]

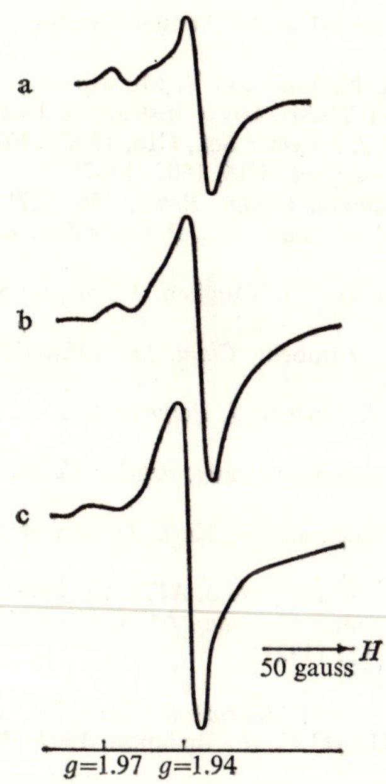

Fig. 2.47. ESR spectra of AA-$TiCl_3$ with alkylaluminium (a, $AlEt_3$; b, $AlEt_2Cl$; c, $AlEtCl_2$).[38]

from the porcelain grinder by the use of a steel ball mill. A similar signal is found with H-grade $TiCl_3$ which is free of $AlCl_3$. The same spectrum also appeared in unground A- and H-grades which were brought into contact with ethylene, propylene or alkylaluminium ($AlEt_3$, $AlEt_2Cl$, $AlEtCl_2$), as shown in Fig. 2.46 and 2.47. We may therefore attribute this spectrum to a lattice defect formed on the surface of the $TiCl_3$ crystal by elimination of a chlorine atom.

REFERENCES

1. G. Natta and I. Pasquon, *Chim. Ind.* (*Milan*), **41**, 993 (1959).
2. K. Soga and T. Keii, *J. Polymer Sci.*, part **A1**, **4**, 2429 (1966).
3. G. Natta, G. Mazzanti, D. De Luca, U. Giannini and F. Bandini, *Makromol. Chem.*, **71**, 54 (1964).
4. G. Natta and I. Pasquon, *Advan. Catalysis*, **11**, 1 (1959).
5. G. Natta, I. Pasquon and E. Giachetti, *Angew. Chem.*, **69**, 213 (1957).
6. G. Natta, I. Pasquon and E. Giachetti, *Chem. Ind.* (*Milan*), **39**, 1002 (1957).
7. T. Keii, *Nature*, **196**, 160 (1962).
8. T, Keii, M. Taira and T. Takagi, presented at the Annual Meeting of Japan Chem. Soc. (1961).
9. T. Keii, T. Takagi, M. Taira, K. Soga, N. Saiki and Y. Murayama, *Shokubai* (*Tokyo*), **5**, 243 (1963); K. Soga, M. Eng. Thesis, Tokyo Institute of Technology (1965); T. Keii, K. Soga and N. Saiki, *J. Polymer Sci.*, **C16**, 1507 (1967).
10. T. Keii, K. Soga and N. Saiki, *J. Polymer Sci.*, **C16**, 1507 (1967).
11. A. Martinato, Y. Chauvin and G. Lefevre, *Compt. Rend.*, **258**, 4271 (1964).
12. J. Ambroz, P. Osecky, J. Mejzlik and O. Hamrik, *J. Polymer Sci.*, **C16**, 423 (1967).
13. W. Cooper, D.E. Eaves, G.D.T. Owen and G. Vaughan, *J. Polymer Sci.*, **C4**, 211 (1963).
14. C. Natta, I. Pasquon, J. Srab and A. Zambelli, *Chim. Ind.* (*Milan*), **44**, 621 (1962).
15. A.K. Ingberman, I. J. Levine and R.J. Turbett, *J. Polymer Sci.*, part A1, **4**, 2781 (1966).
16. A. Schindler and L. Reich, *Polymerization by Organometallic Compounds*, p. 328, Interscience, 1966.
17. H. Schnecko, M. Reinmoller, K. Weirauch and W. Kern, *J. Polymer Sci.*, **C4**, 71 (1963).
18. G. Twigg and E.K. Rideal, *Proc. Roy. Soc.* (*London*), **A171**, 55 (1939).
19. D.D. Eley and E.K. Rideal, *Proc. Roy. Soc.* (*London*), **A178**, 429 (1941).
20. I. Langmuir, *Trans. Faraday Soc.*, **17**, 621 (1921).
21. K. Vessely, *Pure Appl. Chem.*, **4**, 407 (1962).
22. A. Schindler, *J. Polymer Sci.*, **C4**, 81 (1964); *Makromol. Chem.*, **70**, 94 (1964).
23. P.J.T. Tait *et al.*, presented at the IUPAC Conf., Budapest, 1969. Preprints, vol 2, p. 59.
24. G. Natta, P. Corradini and G. Allegra, *J. Polymer Sci.*, **51**, 399 (1961).
25. *Gmellins Handbuch der anorganischen Chemie*, p. 8, 41, Verlag Chemie, 1951.
26. G. Natta, P. Corradini, I.W. Bassi and L. Porri, *Atti Accad. Naz. Lincei Rend. Classe Sci. Fis. Mat. Nat.*, **26**, 155 (1959).

27. T. Keii, T. Suzuki and K. Soga, *Asahi Glass Kogyogijitsu Shoreikai Report* (Japanese), **12,** 155 (1966).
28. S. Brunauer, P.H. Emmett and E. Teller, *J. Am. Chem. Soc.*, **60**, 309 (1938).
29. T. Tanaka, *Kagaku Kogyo*, **18**, 160 (1954).
30. D. Laws and J. Page, *Soil Sci.*, **13**, 319 (1942).
31. G. Naeser, W. Scholz and A. Fiedler, *Kolloid Z.*, **138**, 147 (1963).
32. G. Naeser, *Archiv fur des Einsenhuttenwesen*, **34**, 871 (1963).
33. Esso Research and Eng. Co., *U.S. Pat.*, 745124, 1958.
34. T. Keii and T. Akiyama, *Nature*, **203**, 76 (1964).
35. D.H. Hoeg and G.E. Ashby, quoted in D.F. Hoeg, *The Stereochemistry of Macromolecules* (ed. A.D. Ketley), vol. 1, p. 47, Dekker, 1967.
36. A.D. Ketley and D. Daniels, *Fr. Pat.* 1377037, 1964.
37. Y. Ono and T. Keii, *J. Polymer Sci.*, **A4**, 2441 (1966).
38. I. Okura, Y. Sendoda and T. Keii, *Kogyo Kagaku Zasshi* (Japanese), **73,** 276 (1970).

Chapter 3

Kinetic models for propylene polymerization

In this chapter, the observed kinetic behavior during the polymerization of propylene with Ziegler-Natta catalysts will be discussed in the light of kinetic theory and some kinetic models will be presented.

3.1 KINETIC MODELS FOR STATIONARY POLYMERIZATION

As was first pointed out by Mark *et al.*,[1] the observed rate laws for propylene polymerization can be understood on the basis of adsorption kinetics, which may generally be discussed in terms of the Langmuir theory of adsorption.

3.1.1 The Langmuir adsorption equation

The Langmuir model is based on the assumptions that (i) adsorption of a molecule or atom takes place on an adsorption site, and only one molecule can be accepted by each site, (ii) the surface sites all have identical heats of adsorption, and (iii) there are no energetic interactions between adsorbed molecules (i.e. the heat of adsorption is independent of the surface coverage). Let the number of adsorption sites be N_S, and N_A the number of adsorbed molecules in equilibrium with a gas pressure P at temperature T. The equilibrium rate of adsorption, V_a, will then be given by

$$V_a = k_a P(N_S - N_A) \tag{3.1}$$

and the equilibrium rate of desorption, V_d, by

$$V_d = k_d N_A \tag{3.2}$$

Since both are equilibrium rates they will be equal,

$$V_a = V_d \tag{3.3}$$

and we therefore have

$$N_A = \frac{N_S\left(\frac{k_a}{k_d}\right)P}{1 + \left(\frac{k_a}{k_d}\right)P} \tag{3.4}$$

Putting $(k_a/k_d) = K$ and $(N_A/N_S) = \theta$, Eq. 3.4 may be rewritten as

$$\theta = \frac{KP}{1 + KP} \tag{3.5}$$

where θ is the degree of coverage of the surface sites, and K is the adsorption equilibrium constant. For the adsorption of solute molecules from a solution, P may be replaced by C, the concentration of the solute. The adsorption equilibrium constant can be expressed as

$$K = K_0 e^{E/RT} \tag{3.6}$$

where E is the energy of adsorption (the energy difference between the molecule in the gas or liquid phase and in the adsorbed state). Strictly speaking, E differs from the heat of adsorption by RT, but since $E \gg RT$ for chemisorption, we may neglect this distinction in the field of heterogeneous catalysis.

If there are two types of adsorbed molecules or atoms, A and B, we obtain the adsorption equations for an A–B gas mixture.

$$\theta_A = \frac{K_A P_A}{1 + K_A P_A + K_B P_B}$$
$$\theta_B = \frac{K_B P_B}{1 + K_A P_A + K_B P_B} \tag{3.7}$$

If gaseous molecules such as H_2 are adsorbed in a dissociated form, H, the adsorption equation can be rewritten as

$$\theta_{\mathrm{H}} = \frac{(KP_{\mathrm{H_2}})^{1/2}}{1 + (KP_{\mathrm{H_2}})^{1/2}} \tag{3.8}$$

with

$$K = K_0 e^{E_{\mathrm{H_2}}/RT}$$

where $E_{\mathrm{H_2}}$ represents the heat of adsorption corresponding to $H_2 \rightarrow 2H_{ads}$.

3.1.2 Rideal and Langmuir-Hinshelwood rate laws

Rideal and Langmuir-Hinshelwood type rate laws can be derived from two simple kinetic models of surface reactions. The Rideal model is applicable for the surface reaction of an adsorbed atom or molecule, A, with a gaseous (or solute) molecule, B. In this case the rate of surface reaction will be

$$R = k_s P_{\mathrm{B}} \theta_{\mathrm{A}} \tag{3.9}$$

and substituting for θ_{A} from Eq. 3.5, we obtain

$$R = k_s P_{\mathrm{B}} \left(\frac{K_{\mathrm{A}} P_{\mathrm{A}}}{1 + K_{\mathrm{A}} P_{\mathrm{A}}} \right) \tag{3.10}$$

which is a Rideal rate law.

The Langmuir-Hinshelwood model derives from a consideration of the reaction of two different adsorbed species on the surface. The rate law is found in the same way as for the Rideal model, applying Eq. 3.7 for θ_{A} and θ_{B}.

$$\begin{aligned} R &= k_s \theta_{\mathrm{A}} \theta_{\mathrm{B}} \\ &= k_s \frac{K_{\mathrm{A}} K_{\mathrm{B}} P_{\mathrm{A}} P_{\mathrm{B}}}{(1 + K_{\mathrm{A}} P_{\mathrm{A}} + K_{\mathrm{B}} P_{\mathrm{B}})^2} \end{aligned} \tag{3.11}$$

If the reaction product or some additive, C, is also capable of adsorption, the rate laws derived above will be modified as follows:

$$R = k_s P_{\mathrm{A}} \left(\frac{K_{\mathrm{A}} P_{\mathrm{A}}}{1 + K_{\mathrm{A}} P_{\mathrm{A}} + K_{\mathrm{C}} P_{\mathrm{C}}} \right) \tag{3.12}$$

$$R = k_s \left(\frac{K_{\mathrm{A}} K_{\mathrm{B}} P_{\mathrm{A}} P_{\mathrm{B}}}{(1 + K_{\mathrm{A}} P_{\mathrm{A}} + K_{\mathrm{B}} P_{\mathrm{B}} + K_{\mathrm{C}} P_{\mathrm{C}})^2} \right) \tag{3.13}$$

Many modifications to these two models have been proposed, but with the exception of some models for reactions on independent adsorption sites and on non-uniform surfaces, these are of little value in considering Ziegler-Natta polymerization.

We shall now discuss the Ziegler-Natta polymerization of propylene in terms of a Rideal type model. In this case we may assume that the rate determining step, i.e. the surface reaction, is

$$(\mathrm{AlEt_3\ dimer})_{\mathrm{ads}} + (\mathrm{propylene\ monomer})_{\mathrm{solution}} \xrightarrow{k_s} \mathrm{product} \quad (3.14)$$

The adsorption term ($K[\mathrm{A}]/1 + K[\mathrm{A}]$) in the observed rate law, Eq. 2.11*a*, corresponds to the adsorption coverage of $AlEt_3$ dimer

$$\theta_{\mathrm{Al_2Et_6}} = \frac{K[\mathrm{A}]}{1 + K[\mathrm{A}]} \quad (3.15)$$

At temperatures below 80°C, $AlEt_3$ in solution is completely in the dimer form, but if dissociative adsorption took place on the reaction surface, the coverage of $AlEt_3$ would then be

$$\theta_{\mathrm{AlEt_3}} = \frac{(K[\mathrm{A}])^{1/2}}{1 + (K[\mathrm{A}])^{1/2}} \quad (3.16)$$

as indicated by Eq. 3.8. (It may be noted that the transfer reaction, whose rate is proportional to $[\mathrm{A}]^{1/2}$, has generally been attributed to the action of $AlEt_3$ monomer.) We may derive the stationary polymerization rate (g-C_3H_6/(g-$TiCl_3$ hr atm)) as follows:

$$\begin{aligned} R_\infty &= k_s' N_S \theta_{\mathrm{Al_2Et_6}}[\mathrm{M}] \\ &= k_s HP\left(\frac{K[\mathrm{A}]}{1 + K[\mathrm{A}]}\right) \end{aligned} \quad (3.17)$$

with $E_S - E_{\mathrm{sol}} = 10$ kcal/mole

$E = 12$ kcal/mole

The observed activation energy, which will be equal to $E_S - E_{\mathrm{sol}}$, is 10 kcal/mole. The observed heat of solution, E_{sol}, of propylene in *n*-heptane is 4 kcal/mole,[2)] and the true activation energy of k_s is therefore $E_S = 14$ kcal/mole, as pointed out by Natta *et al.*[2)] The heat of adsorption of Al_2Et_6 from solution is 12 kcal/mole.

On the basis of the Langmuir-Hinshelwood model, the rate-determining step is taken to be

$$(Al_2Et_6)_{ads} + (C_3H_6)_{ads} \xrightarrow{k_s} \quad (3.18)$$

where both species are adsorbed on sites of the same kind. As in Eq. 2.16*c*, provided that $K_M[M] \ll K_A[A] < 1$, we may approximate as follows

$$R_\infty = \left(\frac{k_s H K_M}{2}\right) P \left(\frac{K[A]}{1 + K[A]}\right) \quad (3.19)$$

At higher $AlEt_3$ concentrations in the range $[A] \simeq [A]_{max}$, where $[A]_{max}$ is the concentration at which R_∞ becomes a maximum, Eq. 3.19 may be further approximated to

$$R_\infty \simeq \left(\frac{k_s H K_M}{2}\right) P \quad (3.20)$$

This is similar to the original rate law (Eq. 2.10) proposed by Natta *et al.* In this case, the observed activation energy (10 kcal/mole) should be given by

$$E_S - E_{sol} - E_M = 10 \text{ kcal/mole} \quad (3.21)$$

where E_S is the activation energy of the surface reaction, E_{sol} the heat of solution of propylene, and E_M the heat of adsorption of propylene monomer. The participation of the last term is significant, and a dependence of the polymerization rate on the nature of the monomer is therefore to be expected, according to the Langmuir-Hinshelwood model.

3.1.3 Cossee's model. Surface reaction between independently adsorbed species

Cossee[3)] proposed a model for polymer chain growth on the surface of a $TiCl_3$ catalyst as follows:

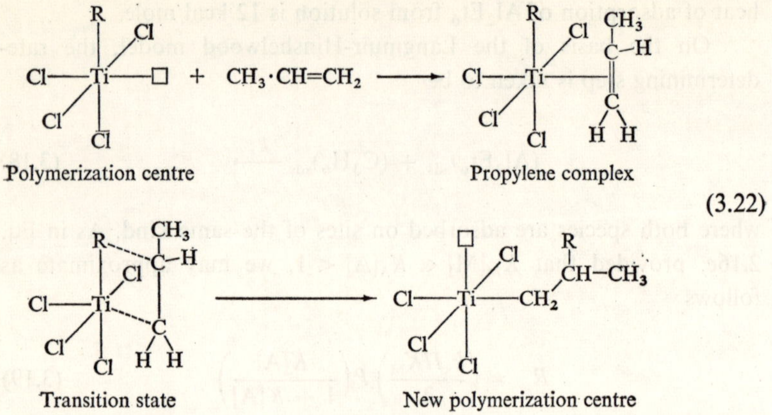

(3.22)

where □ represents an octahedral vacancy, and R is an alkyl group.

This theory is based on quantum chemical considerations of chain growth rather than kinetic data, and it is therefore worthwhile to discuss Cossee's ideas in the light of kinetic theory. The reaction loci are those structures referred to as "polymerization centres", and we may therefore represent the reaction as follows

$$\text{P.C.} + C_3H_6 \underset{\text{adsorption}}{\rightleftharpoons} \text{Propylene complex} \xrightarrow{k_s} \text{P.C.} \qquad (3.23)$$

Transition state

from which the rate law, Eq. 3.24, with respect to propylene may be derived.

$$R_\infty = k_s[\text{P.C.}]\theta_M$$
$$= k_s[\text{P.C.}]\left(\frac{K_M[M]}{1 + K_M[M]}\right) \qquad (3.24)$$

Eq. 3.25 shows the formation of a polymerization centre (P.C.) according to Cossee.

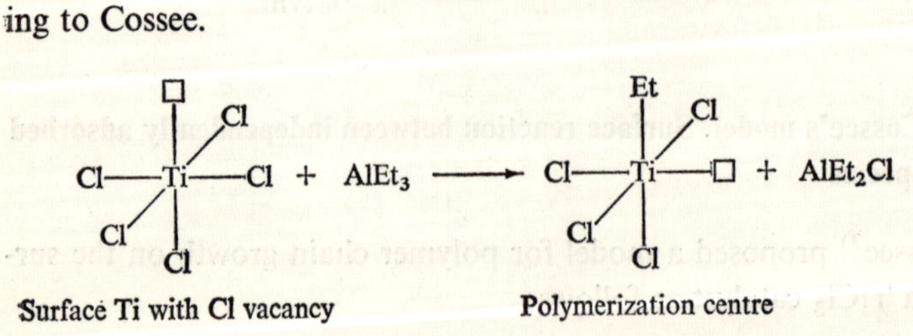

(3.25)

The reaction may thus be a substitution reaction between a surface Ti atom with a chlorine defect and an $AlEt_3$ monomer molecule. At the equilibrium of this reaction, the following equation should hold, from the law of mass action.

$$\frac{[\text{P.C.}][\text{AlEt}_2\text{Cl}]}{[\text{Ti–}\square][\text{AlEt}_3]} = K_A$$

with

$$[\text{Ti–}\square] = [\text{Ti–}\square]_0 - [\text{P.C.}]$$

$$[\text{AlEt}_2\text{Cl}] = [\text{P.C.}]$$

$$[\text{AlEt}_3] \simeq [\text{AlEt}_3]_0$$

Subscript 0 indicates the initial concentrations. Solving these equations, we have

$$[\text{P.C.}] \simeq [\text{Ti–}\square]_0 \tag{3.26}$$

for the case where $[\text{Ti–}\square]_0 \ll K_A[\text{AlEt}_3]_0$, or

$$[\text{P.C.}] \simeq [\text{Ti–}\square]_0\left(\frac{K_A[\text{AlEt}_3]_0}{[\text{Ti–}\square]_0}\right)^{1/2} \tag{3.27}$$

where $[\text{Ti–}\square]_0 \gg K_A[\text{AlEt}_3]_0$.

The concentration of $AlEt_3$ monomer may be expressed as $[\text{AlEt}_3]_0 = (K_D[\text{A}])^{1/2}$, where [A] is the concentration of Al_2Et_6. For the case corresponding to irreversible formation of the polymerization centres, i.e. where $K_A \gg 1$ and $[\text{A}] \gg [\text{Ti–}\square]_0$, as in Eq. 3.26, the rate law given in Eq. 3.24 becomes

$$R_\infty = k_s[\text{Ti–}\square]_0\left(\frac{K_M[\text{M}]}{1 + K_M[\text{M}]}\right) \tag{3.28}$$

On the other hand, where formation of the polymerization centres is reversible, i.e. where $K_A \ll 1$ and $[\text{A}] \ll [\text{Ti–}\square]_0$, Eq. 3.27 can be used to substitute for [P.C.] in Eq. 3.24 to give

$$R_\infty = k_s(K_A^{1/2}K_D^{1/4})[\text{Ti–}\square]_0^{1/2}[\text{A}]^{1/4}\left(\frac{K_M[\text{M}]}{1 + K_M[\text{M}]}\right) \tag{3.29}$$

In practice we would not expect the latter situation to occur, since

$[\text{Ti-}\square]_0/\text{TiCl}_3 < 10^{-2} \sim 10^{-3}$ and $[\text{A}]/[\text{Ti}] > 10^{-2}$ in normal polymerization systems, so that we may expect a rate law such as Eq. 3.28 on the basis of Cossee's model.

This rate law can account for the observations of Natta (Eq. 2.10) provided that $K_M \ll 1$. This means that the observed rate constant $k_s[\text{Ti-}\square]_0(K_M)$ must be smaller than the upper value $k_s[\text{Ti-}\square]_0$. This requirement will now be discussed in the light of the statistical mechanics of adsorption kinetics.

3.2 STATISTICAL MECHANICAL THEORY OF STATIONARY POLYMERIZATION RATES

The stationary rates of propylene polymerization have been discussed by Keii,[4] on the basis of statistical mechanics.[5]

The reaction scheme for propylene polymerization is shown diagrammatically below

$$\left.\begin{array}{ccc} \text{gaseous } C_3H_6 & \rightleftharpoons & \text{monomer in solution} \\ M_g & (\text{I}) & M \\ & & \\ & & \text{Polymerization centre} \\ & & \text{P.C.} \end{array}\right\} \rightleftharpoons \underset{\text{(II)}}{} \ \underset{M^{\neq}}{\text{Activated complex}} \rightarrow \underset{\text{(III)}}{\text{P.C.}} \quad (3.30)$$

Denoting the surface concentration of the activated complex by $[M^{\neq}]$, the rate of polymerization per unit surface area can be expressed as follows, stage III being the rate-determining step:

$$R_\infty = \frac{kT}{h}[M^{\neq}] \quad (3.31)$$

where k is the Boltzmann constant, h is Planck's constant and T(°K) the absolute temperature. In order to apply the theory to the various kinetic models described in the previous section, further development is necessary. Readers familiar with the statistical mechanical approach to kinetics may omit the remainder of this section.

The absolute activity λ_δ of a species δ may be defined as

$$\lambda_\delta = e^{\mu_\delta/RT} \quad (3.32)$$

where μ_δ is the chemical potential of δ. Since all reaction steps other than III are assumed to be in partial equilibrium, the following equations must hold

$$\begin{aligned} \mu_g &= \mu_{\mathrm{M}}, \quad \lambda_g = \lambda_{\mathrm{M}} \\ \mu_{\mathrm{C}} + \mu_{\mathrm{M}} &= \mu^{\neq}, \quad \lambda_{\mathrm{C}}\lambda_{\mathrm{M}} = \lambda^{\neq} \end{aligned} \tag{3.33}$$

where g represents gaseous propylene, M is monomer in solution, C is a polymerization centre and $\neq$ represents the activated complex.

As previously mentioned, Henry's law holds in these polymerization systems, so that the propylene gas and the monomer solution can be regarded as ideal. We can therefore apply the following ideal state equations

$$\begin{aligned} \mu_g &= \mu_g{}^0 + kT \ln [\mathrm{M}_g] \\ &= \mu_g{}^{0\prime} + kT \ln P \\ \mu_{\mathrm{M}} &= \mu_{\mathrm{M}}{}^0 + kT \ln [\mathrm{M}] \end{aligned} \tag{3.34}$$

from which Henry's law can be derived, using Eq. 3.33, as follows

$$[\mathrm{M}] = Pe^{-(\mu_{\mathrm{M}}{}^0 - \mu_g{}^{0\prime})/kT} \tag{3.35}$$

Using Eq. 3.34, the absolute activities λ_g and λ_{M} can be written as

$$\lambda_g = e^{\mu_g/kT} = \frac{[\mathrm{M}_g]}{Q_g} \tag{3.36}$$

$$\lambda_{\mathrm{M}} = e^{\mu_{\mathrm{M}}/kT} = \frac{[\mathrm{M}]}{Q_{\mathrm{M}}} \tag{3.37}$$

with

$$\begin{aligned} Q_g &= e^{-\mu_g{}^0/kT} \\ Q_{\mathrm{M}} &= e^{-\mu_{\mathrm{M}}{}^0/kT} \end{aligned} \tag{3.38}$$

The function Q_g is the molecular partition function of gaseous propylene, which can be further expressed as the product of the translational, rotational and vibrational partition functions, Q_{tr}, Q_{rot} and Q_{vib}, respectively.

$$Q_g = Q_{\mathrm{tr}} Q_{\mathrm{rot}} Q_{\mathrm{vib}} e^{-E_g{}^0/kT} \tag{3.39}$$

where $E_g{}^0$ is the ground state energy of the gaseous propylene molecule. The absolute activity of an ideal gas or solution is proportional to the concentration, as can be seen from Eq. 3.36 and 3.37, but the surface activities are not proportional to the surface concentrations except at low values of surface concentration. The general equation for absolute activity has been formulated by Horiuti[5] as

$$\lambda_\delta = \frac{\theta_{\sigma(\delta)}}{q_\delta \theta_{\sigma(0)}} \tag{3.40}$$

where $\theta_{\sigma(\delta)}$ and $\theta_{\sigma(0)}$ represent the probability that the site σ is occupied by δ or vacant, and q_δ is the partition function of δ in σ. As the spatial extension of the site tends to zero, $q_\delta Q_\delta$ and $\theta_{\sigma(0)} \rightarrow V$, the volume, and $\mathcal{L}_{\sigma(\delta)} \rightarrow N_\delta$. The general form of Eq. 3.40 then reduces to Eq. 3.36 and 3.37. In the case of surface reactions, σ is a surface site, and $\theta_{\sigma(\delta)}$ is then the coverage, θ, of the sites by δ. $\theta_{\sigma(0)}$ can then be expressed as $(1 - \theta)$, and Eq. 3.40 reduces to a Langmuir adsorption equation.

$$\frac{\theta}{1-\theta} = q_\delta \lambda_\delta \tag{3.41}$$

with

$$\lambda_\delta = \frac{P}{kTQ_\delta}$$

The general form, Eq. 3.40, can be applied to an activated complex on a surface in the form

$$\lambda^{\neq} = \frac{\theta_{\sigma^{\neq}(M^{\neq})}}{q^{\neq}\theta_{\sigma^{\neq}(0)}} \tag{3.42}$$

where $\sigma^{\neq}$ is a surface site available for an activated complex $M^{\neq}$. Denoting the surface concentration of such sites by $G^{\neq}$, the surface concentration of activated complex can be expressed as

$$\begin{aligned}[M^{\neq}] &= G^{\neq}\theta_{\sigma^{\neq}(M^{\neq})} \\ &= G^{\neq}q^{\neq}\theta_{\sigma^{\neq}(0)}\lambda^{\neq}\end{aligned} \tag{3.43}$$

Substituting this expression for $[M^{\neq}]$ into Eq. 3.31, and applying Eq. 3.33, we obtain

$$R_\infty = \frac{kT}{h} G^{\neq}\theta_{\sigma^{\neq}(0)} q^{\neq} \lambda_C \lambda_g \tag{3.44}$$

Further deductions can now be made based on particular kinetic models.

3.2.1 The Rideal model

In the Rideal model, we may assume that the activated complex is of the form shown in Eq. 3.45

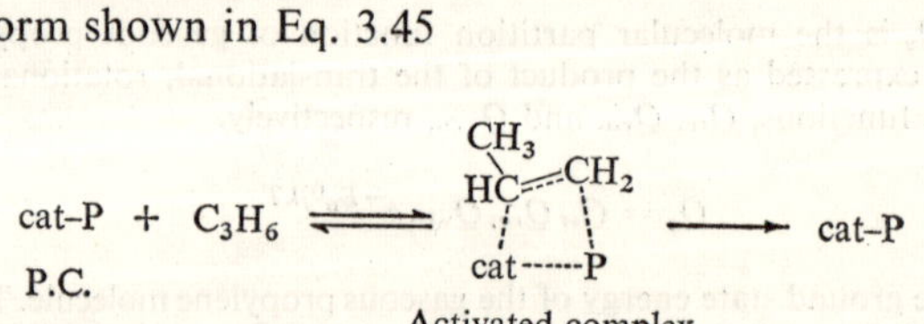

(3.45)

The partition function of the activated complex, $q^{\neq}$, can be derived if it is assumed that the complex involves an alkyl group, P, and one propylene monomer unit which has lost all degrees of translational and rotational freedom. The motion of the alkyl group, P, is assumed to be the same both in the activated complex and in the polymerization centre, though P will be in a higher energy state in the former case. We then have

$$q^{\neq} = q_C Q_{vib} e^{-E^{\neq}/kT} \tag{3.46}$$

where q_C is the partition function of the polymerization centre (cat-P). From Eq. 3.41, the absolute activity, λ_C, of the polymerization centre may be expressed as

$$\lambda_C = \frac{\theta_{\sigma(C)}}{q_C \theta_{\sigma(0)}} \tag{3.47}$$

and substituting from Eq. 3.36, 3.39, 3.46 and 3.47 into Eq. 3.44, we obtain the following expression for the stationary rate of polymerization.

$$R_\infty = \frac{kT}{h} G_\sigma \theta_{\sigma(C)} \frac{e^{-(E^{\neq} - E_g{}^0)/kT}}{Q_{tr} Q_{rot}} [M_g] \tag{3.48}$$

The translational partition function of gaseous propylene is given by

$$Q_{tr} = \frac{(2\pi mkT)^{3/2}}{h^3} \tag{3.49}$$

and substituting into Eq. 3.48 we can derive

$$R_\infty = \frac{PS}{(2\pi M/RT)^{1/2}} \frac{(G_\sigma/S)}{(2\pi mkT/h^2)} \frac{\theta_{\sigma(C)}}{Q_{rot}} e^{-E/RT} \tag{3.50}$$

where S is the surface area of the $TiCl_3$ surface in Eq. 3.50, M is the molecular weight of propylene and $E = E^{\neq} - E_g^0$. The first factor in Eq. 3.50 is the so-called "hitting number" of molecules on the surface S, and the remaining factor is the "sticking probability."

We may rewrite the second factor in Eq. 3.50 as follows

$$\frac{(G_\sigma/S)}{(2\pi mkT/h^2)} = \frac{1}{(S/G_\sigma)(2\pi mkT/h^2)} \tag{3.51}$$

The translational partition function of a molecule having two dimensional translational motion within a small area, s', can be expressed as follows,

$$Q_{g,s'} = s' \frac{2\pi mkT}{h^2} \tag{3.52}$$

and, when s' is a molecular dimension, this can be regarded as a two dimensional vibrational partition function. In this case the value of $s'(= S/G_\sigma)$ may be less than 10Å^2, since $G_\sigma/S \simeq 10^{15}$ cm^{-2}. We therefore have the approximation†

$$\frac{S}{G_\sigma}\left(\frac{2\pi mkT}{h^2}\right) \simeq 1 \tag{3.53}$$

If the formation of the polymerization centre is assumed to be due to the adsorption of Al_2Et_6, then the function $\theta_{\sigma(C)}$ becomes

$$\theta_{\sigma(C)} = \frac{\lambda_A q_A}{1 + \lambda_A q_A} = \frac{K[A]}{1 + K[A]} \tag{3.54}$$

with $$K = q_A/Q_A$$

The rate of polymerization can then be expressed as

$$R_\infty = (Z_0 S)\frac{1}{Q_{rot}}\left(\frac{K[A]}{1 + K[A]}\right)e^{-E/RT} \tag{3.55}$$

where $Z_0 S$ is the "surface hitting number" at propylene pressure P and temperature T. We can put E equal to the activation energy (10 kcal/mole), referred to gaseous propylene. According to the experimental data of Keii *et al.*,[6] the stationary rate of polymerization is 0.4 g-C_3H_6/(m^2 hr) at 43°C and 66 cm Hg. The rates reported by most authors fall in the range 10 ~ 40 g-C_3H_6/(g-$TiCl_3$ hr atm), corresponding to 0.5 ~ 1 g-C_3H_6/(m^2 hr atm) at room temperature. All these data give about $3 \times 10^{21} \sim 10^{22}$ molecules/(cm^2 sec atm) for the pre-exponential factor A of the experimental rate law (Eq. 3.56) in the region where it is independent of [A].

$$R_\infty = Ae^{-(10\text{kcal/mole})/RT} \qquad P = 1 \text{ atm}$$

$$A = \frac{(0.5 \sim 1.0) \times 10^7 \times 6 \times 10^{23}}{42 \times 3600 \times 10^4} \tag{3.56}$$

$$= 2 \sim 4 \times 10^{21} \text{ molecules/(cm}^2\text{ sec atm)}$$

† I would like to note here the kind criticism of Prof. Boreskov on this point after my lecture at Novosibirsk in 1968.

The theoretical value of the pre-exponential factor as calculated from Eq. 3.55 for the case where $\theta_{\sigma(C)} = 1$ (i.e. $K[A]/(1 + K[A]) = 1$) is about 3×10^{21} molecules/(cm^2 sec), using $Z_0 = 2.54 \times 10^{23}$ molecules/(cm^2 sec)$^{-1}$ at 25°C† and $Q_{rot} \simeq 100$. The agreement is remarkably good. It should be remembered that the calculated value is based on values of G_σ approaching the upper limit. We can conclude that the Rideal model is appropriate for propylene polymerization with Ziegler-Natta catalysts, and that all the Ti surface atoms participate in the polymerization as active centres.

3.2.2 The Langmuir-Hinshelwood model

A physical model suitable for Langmuir-Hinshelwood polymerization is shown below

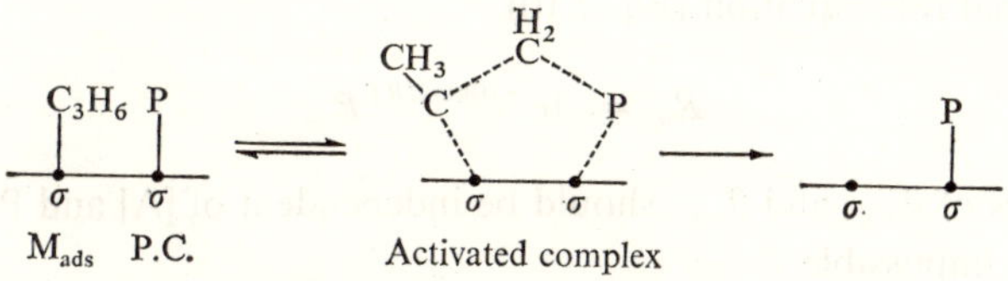

(3.57)

In this case, the following relationships hold

$$\theta_{\sigma^{\neq}(0)} = \theta^2_{\sigma(0)} = (1 - \theta_{\sigma(M)} - \theta_{\sigma(C)})^2 \tag{3.58}$$

$$\theta_{\sigma(0)} = \frac{1}{1 + \lambda_g q_M + \lambda_A q_C} = \frac{1}{1 + K_M[M] + K_A[A]} \tag{3.59}$$

$$\lambda_C = \frac{\theta_{\sigma(C)}}{q_C(1 - \theta_{\sigma(M)} - \theta_{\sigma(C)})} \tag{3.60}$$

$$\theta_{\sigma(C)} = \frac{K_A[A]}{1 + K_M[M] + K_A[A]} \tag{3.61}$$

$$\lambda_g = \frac{[M_g]}{Q_g} = \frac{\theta_{\sigma(M)}}{q_M \theta_{\sigma(0)}} \tag{3.62}$$

and from Eq. 3.44 we can derive the rate equation given in Eq. 3.63,

† Calculated for [M] = 42 g/mole by E.A. Guggenheim and J.E. Prue, *Physicochemical Calculation*, p. 58, North-Holland, 1955.

in a similar way to that used for the Rideal model.

$$R_\infty = \frac{kT}{h} G^{\neq} \theta_{\sigma(C)} \theta_{\sigma(0)} \frac{q^{\neq}}{q_C Q_g} e^{-E^{\neq}-E_g{}^0/RT} [M_g] \tag{3.63}$$

Again applying the method used for the Rideal model, the upper limit of R_∞ can be shown to be

$$R_\infty = (Z_0 S) \frac{1}{Q_{rot}} \frac{K_A[A]}{(1 + K_M[M] + K_A[A])^2} e^{-E/RT} \tag{3.64}$$

using the approximation $G^{\neq} = G_\sigma$. This equation gives a good value for the pre-exponential factor subject to the restriction that $K_M[A]/(1 + K_M[M] + K_A[A])^2 \simeq 1$. However, this requires that $\theta_{\sigma(C)} = 1$ and $\theta_{\sigma(0)} = 1$ simultaneously, which cannot be the case. In order to explain the original rate equation (Eq. 2.10)

$$R_\infty = Ae^{-10\text{kcal}/RT} P$$

the values of $\theta_{\sigma(C)}$ and $\theta_{\sigma(0)}$ should be independent of [A] and P, which is clearly impossible.

If the condition $1 \gg K_M[M] + K_A[A]$ is taken, we have $\theta_{\sigma(0)} \simeq 1$ and $\theta_{\sigma(C)} = K_A[A] \ll 1$, which gives rise to a second order rate law $R_\infty \propto [A]P$, with a small pre-exponential factor. The condition $K_A[A] \gg 1 + K_M[M]$ gives $\theta_{\sigma(C)}\theta_{\sigma(0)} \simeq 1/K_A[A] \ll 1$, which results in a rate law of the form $R_\infty \propto P/[A]$, again with a small pre-exponential factor.

3.2.3 Cossee's model

The physical nature of Cossee's model is shown below.
For this model, the following equations will hold

$$\sigma^{\neq} = \sigma, \quad \theta_{\sigma^{\neq}(0)} = \theta_{\sigma(0)} = 1 - \theta_{\sigma(M)}$$

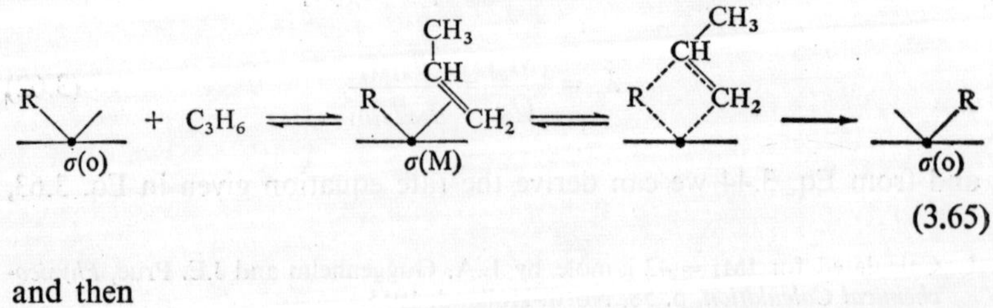

(3.65)

and then

$$R_\infty = \frac{kT}{h} G_\sigma \theta_{\sigma(0)} \frac{q^{\neq}}{Q_g} e^{-E^{\neq} - E_g{}^0/RT}[M_g]$$

$$= (Z_0 S) \frac{1}{Q_{\text{rot}}} \theta_{\sigma(0)} e^{-E/RT} \qquad (3.66)$$

which will give a good value for the pre-exponential factor when $\theta_{\sigma(0)} \simeq 1$, i.e. when $K_M[M] \ll 1$, which corresponds to the case of negligible adsorption of monomer on the polymerization centres. Thus, Cossee's model can tell us no more than the Rideal model about the adsorbed monomer, from a kinetic viewpoint.

3.3 KINETIC MODELS FOR THE BUILD-UP PERIOD

In principle, any kind of kinetic curve can be explained in terms of changes in the number of polymerization centres. In this section, the build-up periods of acceleration and decay type kinetic curves will be discussed in the light of this.

Comparison of the initial rate law obtained with decay type kinetic curves (Eq. 2.5*d* and 2.5*e*) with that derived by the present author for acceleration type curves (Eq. 2.4) suggests that the same kinetic model may be applicable for the build-up periods of both types of curve. We may suppose that the polymerization centres are formed in two stages: (i) the formation of an active site S* from a $TiCl_3$ surface site S by Al_2Et_6, followed by (ii) the formation of a polymerization centre C* from an active site and a propylene monomer unit.

(i) $$S + Al_2Et_6 \rightleftharpoons S^* \qquad (3.67)$$

(ii) $$S^* + C_3H_6 \xrightarrow{k_{C^*}} C^* \qquad (3.68)$$

3.3.1 Acceleration type kinetic curves

The experimental results of Natta *et al.*[2)] on acceleration type kinetic curves were carried out at high concentrations of $AlEt_3$ and from these we can derive information on the dependence of the curves on $AlEt_3$ concentration. For a discussion of Natta's results we shall therefore consider only step (ii) in the formation of polymerization centres.

Denoting the total number of active sites by $[S^*]_0$ and assuming that step (ii) represents an irreversible reaction, the rate of formation of polymerization centres can be expressed as

$$\frac{d[C^*]}{dt} = k_{C*}[M]([S^*]_0 - [C^*]) \tag{3.69}$$

where [M] is the monomer concentration. The polymerization rate is proportional to the number of polymerization centres, i.e. $R = k_p[M][C^*]$ in general, and providing that $k \equiv k_{C*}[M]$ and $R_\infty \equiv k_p[M][S^*]_0$, Eq. 3.69 can be put in the form

$$\frac{dR}{dt} = k(R_\infty - R) \tag{3.70}$$

which is the same as the empirical formula given in Eq. 2.2*a*. According to the results of Natta *et al.*

$$k = \frac{R_\infty}{12} \quad hr^{-1}$$

with $\quad R_\infty = 2 \times 10^7 e^{-10kcal/RT} P$ g-C_3H_6/(g-$TiCl_3$ hr atm)

from which the following relation between the constants k and k_{C*} can be obtained.

$$k = k_{C*}[M]$$

$$= k_{C*}HP \tag{3.71}$$

and $$k_{C*} = A'_{C*}e^{-14kcal/RT} \tag{3.72}$$

$$k_{C*}H = A_{C*}e^{-10kcal/RT} \tag{3.73}$$

The value of the pre-exponential factor A_{C*} could be estimated from Eq. 2.2*d*, 2.2*e* and 2.3, i.e.

$$A_{C*} \simeq \frac{A}{12} \tag{3.74}$$

$$= 10^6\ hr^{-1} \tag{3.75}$$

From the result that the activation energy of k_{C*} is the same as that of k_p we may conclude that step (ii) represents the insertion of a propylene monomer unit into the catalyst-C_2H_5 bond to form the polymerization centre,

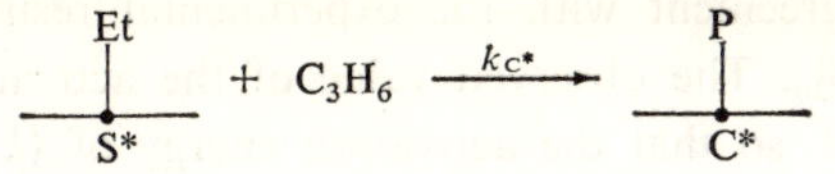

(3.76)

Active centre

and from Eq. 3.74 it can be concluded that the rate of this step is slower than that of monomer insertion into a growing polymer chain, probably due to an entropy factor. The structure of the active centre shown above may be similar to that favoured by Cossee (shown in Eq. 3.25).

3.3.2 The build-up period during decay-type polymerization

As described in (2.2.2), kinetic curves during this period are dependent on the order of addition of propylene and $AlEt_3$ to $TiCl_3$. It was considered by Keii and Soga[7,8] that equilibrium in step (i) is established before the addition of propylene for the usual order of addition (case A, $[TiCl_3 + AlEt_3] + C_2H_6$). These authors gave Eq. 3.77 for the total number of active sites

$$[S^*]_0 = [S]_0\left(\frac{K[A]}{1 + K[A]}\right) \tag{3.77}$$

where $[S]_0$ is the total number of surface sites available for active site formation, and K is the equilibrium constant for step (i).

In the same way as before, we have

$$\frac{d[C^*]}{dt} = k_1'[M]([S^*]_0 - [C^*]) \tag{3.78}$$

and

$$R = \bar{R}_0(1 - e^{-k_1'[M]t}) \tag{3.79}$$

with

$$\bar{R}_0 = k_p[M][S^*]_0$$

$$= k_p[M][S]_0\left(\frac{K[A]}{1 + K[A]}\right) \tag{3.80}$$

At the beginning of the build-up period, where $t \simeq 0$, Eq. 3.79 may be approximated as follows

$$R \simeq \bar{R}_0 k_1'[M]t$$

$$= k_p k_1' H^2 [S]_0 P^2 \left(\frac{K[A]}{1 + K[A]}\right)t \tag{3.81}$$

which is in agreement with the experimental results provided that $k_1 = k_p k_1' H^2[\mathrm{S}]_0$. The observed value of the activation energy of k_1 is 20 kcal/mole, so that the activation energy of $(k_1'H)$ must be the same as that of (k_pH), i.e. 10 kcal/mole. The mechanism of the build-up period can thus be regarded as the same for both types of kinetic curve.

However, the observed values of k_1 may be as much as 3000 times greater than the value of k in Natta's acceleration curves. Thus from Fig. 2.6 we have

$$k_1 = 800\left(\frac{\text{g-C}_3\text{H}_6}{\text{g-TiCl}_3\ \text{hr}^2\ \text{atm}^2}\right) \text{at } 44\,^\circ\text{C} \tag{3.82}$$

whereas Fig. 2.3 shows that $k < 1\,\text{g-C}_3\text{H}_6/(\text{g-TiCl}_3\ \text{hr}^2\ \text{atm}^2)$ at 43 °C. It should also be noted that R_0 represents the final value of the polymerization rate at the end of the build-up period, and is not the same as the stationary rate R_∞.

In case B, where the order of addition was $(\mathrm{TiCl_3} + \mathrm{C_3H_6}) + \mathrm{AlEt_3}$, we may assume the following kinetic model

(i) $$\mathrm{S} + \mathrm{C_3H_6} \rightleftharpoons \mathrm{S}^{*\prime} \tag{3.83}$$

(ii) $$\mathrm{S}^{*\prime} + \mathrm{Al_2Et_6} \xrightarrow{k_2'} \mathrm{C}^* \tag{3.84}$$

In this case, step (ii) will be rate determining, and we therefore have the following equations

$$\frac{d[\mathrm{C}^*]}{dt} = k_2'[\mathrm{A}]([\mathrm{S}^{*\prime}]_0 - [\mathrm{C}^*]) \tag{3.85}$$

$$R = \bar{R}_0'(1 - e^{-k_2'[\mathrm{A}]t}) \tag{3.86}$$

The initial rate, R_0, will be

$$R_0 = k_2' k_p H[\mathrm{S}^{*\prime}]_0 P[\mathrm{A}]t \tag{3.87}$$

which is in agreement with the experimental formula, Eq. 2.6*a*, providing $k_2 \equiv k_2' k_p H[\mathrm{S}^*]_0$. In this case we obtain Eq. 3.88.

$$R_0 = k_2 P[\mathrm{A}]t \tag{3.88}$$

The activation energy of k_2 has been found experimentally to be 17 kcal/mole, and allowing for the 10 kcal/mole contribution from k_pH, the activation energy of k'_2 should therefore be 7 kcal/mole. The value of R_0/t is 2000 g-C_3H_6/(g-$TiCl_3$ hr^2) at $P = 66$ cm Hg, [A] = 15 mmole/l and $T = 43$°C in this case, whereas for case A, R_0/t is 600 g-C_3H_6/(g-$TiCl_3$ hr^2) under the same polymerization conditions.

From Eq. 3.87 and 3.81 the theoretical values of R_0/t in the two cases are

Case B $\quad R_0/t = k'_2[A][S^{*\prime}]_0(k_pHP)$

Case A $\quad R_0/t = k'_1[M][S^*]_0(k_pHP)$

when
$$\frac{K_A[A]}{1 + K_A[A]} = 1$$

The equilibrium concentration of propylene monomer at $P = 1$ atm is 0.5 mole/l[2)] at 44°C, so comparing the two values of R_0/t at the same concentrations [M] and [A] we have

$$\frac{k'_2[S^{*\prime}]_0}{k'_1[S^*]_0} = \frac{2000 \times (1/0.015)}{600 \times (1/0.45)} \simeq 10^2 \tag{3.89}$$

Taking into account the activation energies of k'_2 (7 kcal/mole) and k'_1 (14 kcal/mole), the pre-exponential factors can be compared as follows

$$\frac{A'_2[S^{*\prime}]_0}{A'_1[S^*]_0} \simeq 10^{-3} \tag{3.90}$$

This rather small ratio suggests that the formation of polymerization centres by attack of Al_2Et_6 on the active sites (case B) has both small activation entropy and small activation energy, since

$$\frac{[S^{*\prime}]_0}{[S^*]_0} = \frac{k_pHP[S^{*\prime}]_0}{k_pHP[S^*]_0} = \frac{\bar{R}'_0}{\bar{R}_0} \simeq \frac{140}{115} \tag{3.91}$$

The last ratio is the ratio of the maximum rates in the two kinetic curves (Fig. 2.6 and 2.10). On the basis of the above discussion the mechanisms of steps (ii) and (ii′) may be as follows

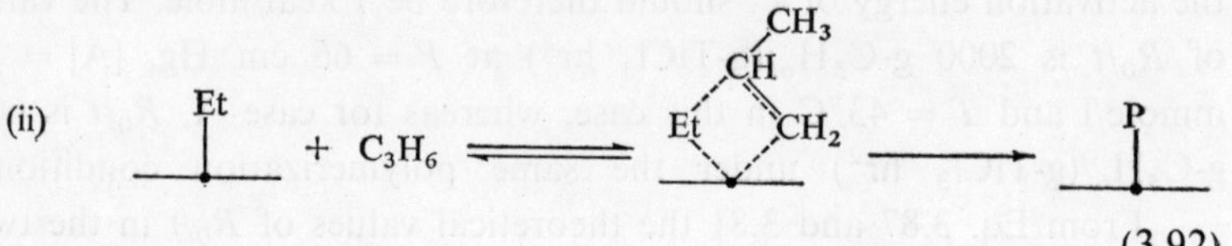

(3.92)

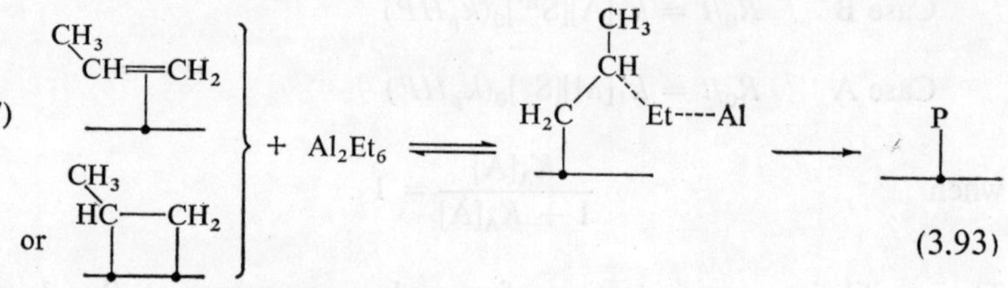

(3.93)

The σ-bonded adsorption structure may be indicated by the experimental fact that propylene polymerization takes place only on $TiCl_3$.

3.4 THE DECAY PERIOD DURING DECAY TYPE POLYMERIZATION

The observation that, during the polymerization of propylene with active $TiCl_3/AlEt_3$ catalysts, the polymerization rate decreases gradually from a maximum down to the stationary rate has been discussed by Natta *et al.*[2] in terms of the agglomeration of $TiCl_3$ particles by the mechanical action of the polymer formed during the reaction. Later they noted the effects of mass transfer on the polymerization under conditions where polymer production exceeded 100 g/l. Fukui *et al.*[9] also considered the effects of polymer formation. The kinetic data applicable to the decay period have been summarized by Keii and Soga[7,8] as follows

$$-\frac{dR}{dt} = k_3(R - R_\infty) \tag{2.7c}$$

$$\frac{R - R_\infty}{R_0 - R_\infty} = e^{-k_3 t} \tag{2.7a}$$

and $$k_3 = k'_{3,0}e^{-(2\sim3)\text{kcal}/RT} + k''_{3,0}e^{-12\text{kcal}/RT}P \tag{2.9}$$

with $$k_3 \simeq k'_3 \gg k''_3P \quad \text{at} \quad P < 1 \text{ atm} \quad \text{and} \quad T \leqslant 80°\text{C} \tag{3.94}$$

The arguments of Natta *et al.* require that $k_3 \propto$ amount of polymer produced $\propto P$. It can be seen from Eq. 3.94 that this is not the case, and the work of Natta *et al.*[10] themselves on propylene polymerization with $VOCl_3/Al(i\text{-}Bu)_3$ indicates that the rate fall-off cannot be due to polymer formation.

Pasquon *et al.*[11] proposed Eq. 3.95 for propylene polymerization with the above catalyst, on the assumption that monomer transfer through the polymer layer covering the catalyst surface was the rate determining step.

$$\frac{1}{R^2} = \frac{kL^2}{G^2[\text{M}]}t + \frac{1}{R_0^2} \tag{3.95}$$

R_0 is the rate at $t = 0$, G is the mass of catalyst, L is a shape factor and [M] the monomer concentration. Keii and Soga[8] compared their experimental results with Eq. 3.95, as shown in Fig. 3.1. The poisoning effect of $AlEtCl_2$ or $AlEt_2Cl$, formed from $TiCl_3$ and $AlEt_3$, has been considered by Caunt[12] and Schnecko *et al.*[13] However, the kinetic

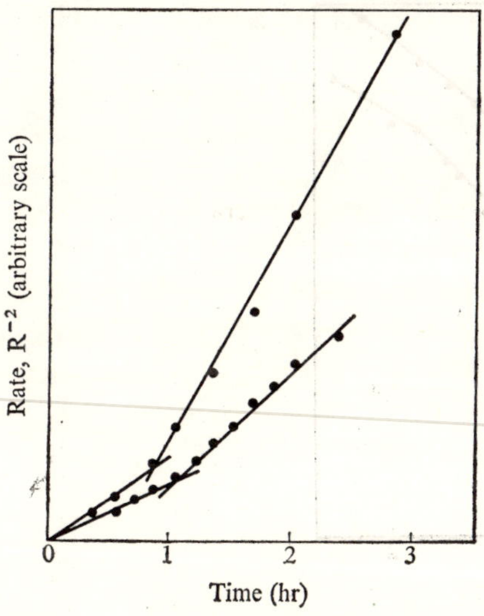

Fig. 3.1. Comparison of Eq. 3.95 with the experimental data of Keii and Soga.[8]

curves obtained during the decay period are independent of $AlEt_3$ concentration according to Smith *et al.*,[14] who favoured a second order deactivation of the polymerization centres, as in Eq. 3.96.

$$-\frac{d[C^*]}{dt} = k[C^*]^2 \tag{3.96}$$

This gives a rate equation of the type

$$\frac{1}{R} - \frac{1}{R_0} = kt \tag{3.97}$$

Fig. 3.2 compares Eq. 3.97 with the experimental results obtained by Keii and Soga.[8] This equation seems preferable to that of Pasquon *et al.* (Eq. 3.95). De Vries[15] found some decomposition of $TiCl_3$/alkyl in homogeneous systems as follows

$$2TiCl_3\cdot alkyl \rightarrow 2\ TiCl_3 + alkane + olefin \tag{3.98}$$

which could account for a second order deactivation of the surface. However, several types of first order mechanism have also been pro-

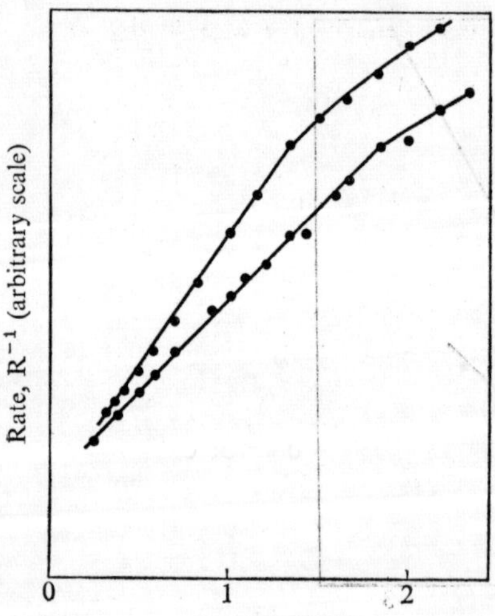

Fig. 3.2. Comparison of Eq. 3.97 with the experimental data of Keii and Soga.[8]

posed, e.g. that of Kohn *et al.*,[16] arising from work on the polymerization of propylene with $TiCl_3/AlEt_2Cl$.

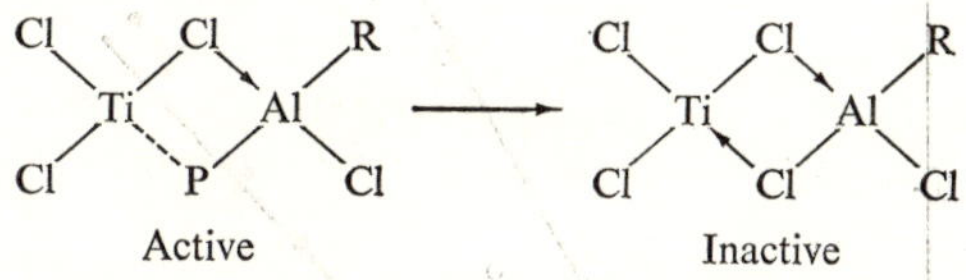

(3.99)

This model has had some success in accounting for the experimental rate laws (Eq. 2.7*a*, 2.7*c* and 2.9).

Ambroz *et al.*[17] examined the system $TiCl_3/AlEt_3/n$-heptane and found that a considerable amount of Cl^- was extracted from $TiCl_3$ into the solution. The rate of extraction could be described by Eq. 3.100

$$-\frac{d[Cl^-]}{dt} = k'([Cl^-]_\infty - [Cl^-]) \tag{3.101}$$

which, on integration, becomes

$$\frac{[Cl^-] - [Cl^-]_\infty}{[Cl^-]_0 - [Cl^-]_\infty} = e^{-k't} \tag{3.100}$$

At the same time, they found similar rate laws for the active centres from the decreasing rate of polymerization.

$$-\frac{d[C^*]}{dt} = k^*([C^*] - [C^*]_\infty) \tag{3.102}$$

or

$$\frac{[C^*] - [C^*]_\infty}{[C^*]_0 - [C^*]_\infty} = e^{-k^*t}$$

The applicability of Eq. 3.100 and 3.102 is shown in Fig. 3.3 for their results. Although the values of k' and k^* do not coincide, the forms of the rate laws are the same, and are in agreement with those found by Keii and Soga (Eq. 2.7*a* and 2.7*b*). This constitutes strong evidence for their suggestion that Cl^- extraction

$$(TiCl_3)_m + (AlEt_3)_n \rightleftharpoons Ti_mCl_{3m-x}Et_x(AlEt_3)_{n-x} + (AlEt_2Cl)_x \tag{3.103}$$

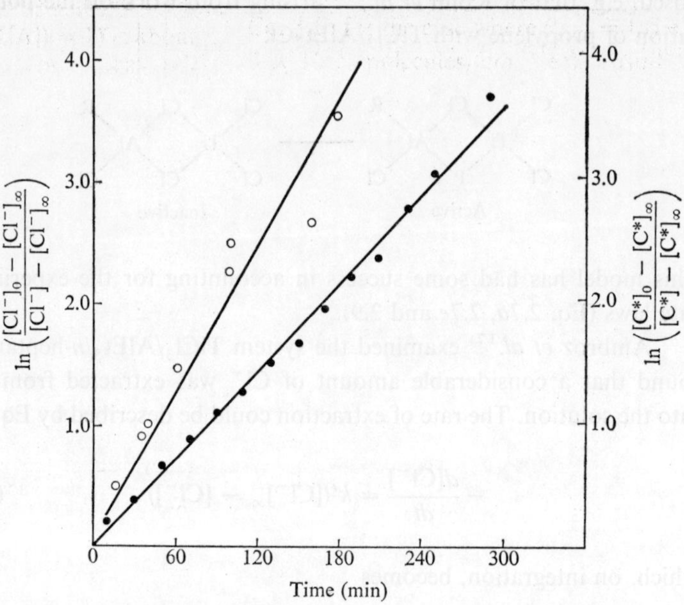

Fig. 3.3. Experimental validity of Eq. 3.100 (○) and 3.102 (●). $k = 2.0 \times 10^{-2}$ min^{-1} or 1.2×10^{-2} respectively.[17]

takes place during polymerization, resulting in the deactivation of some of the active centres. The main difficulty arising from this mechanism lies in the apparent contradiction with the experimental results of Ambroz *et al.*[17] that the kinetic curves are independent of $AlEt_3$ concentration. However, when $TiCl_3$ which had been aged in concentrated $AlEt_3$ solution ($[A]/[Ti] \geq 10$) was used, a substantial lowering of the kinetic curve was found, depending on the aging period. This effect is shown in Fig. 3.4, and has been confirmed by Keii *et al.*,[18] whose results are given in Fig. 3.5.

Keii *et al.*[19] suggested that there were two types of polymerization centres present

(A) Ti–P and (B) 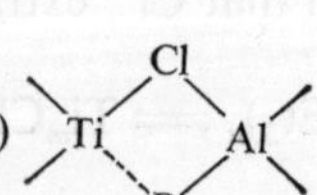 (3.104)

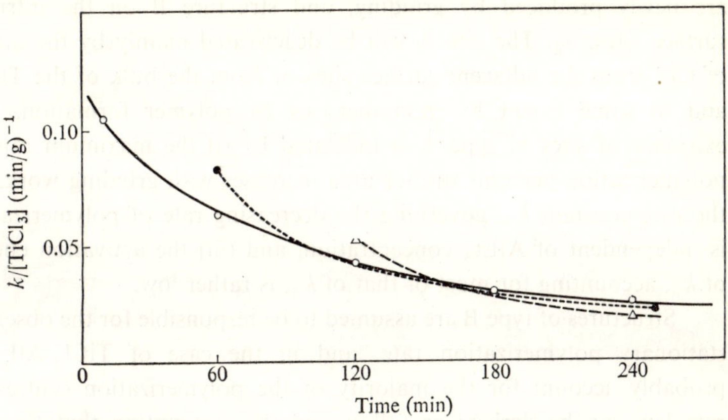

Fig. 3.4. Kinetic curves obtained after various catalyst ageing times. (○ = initial curve, ● = monomer added after 60 min, △ = monomer added after 120 min. $[M]_0 = 2.37$ mole/l, $[AlEt_3] = 5.5 \times 10^{-3}$ mole/l, $[TiCl_3] = 0.05$ g/l, specific surface area 80 m^2/g at 50°C.)

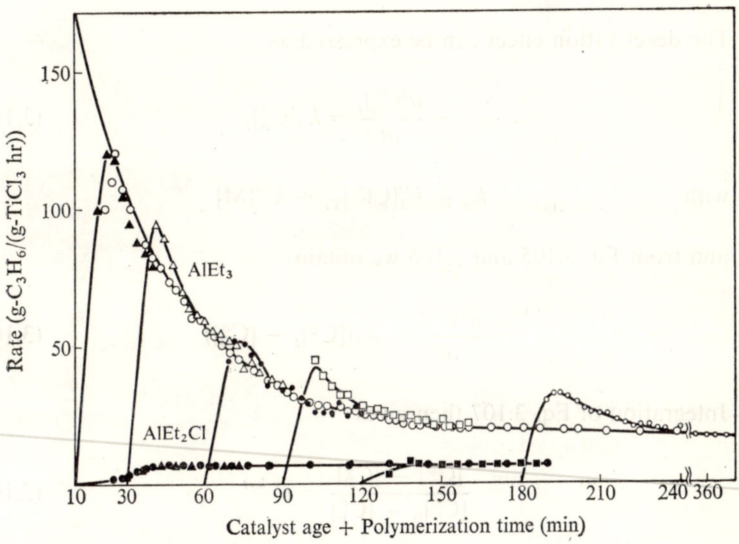

Fig. 3.5. Kinetic curves obtained after various catalyst ageing times. (TAC-$TiCl_3$ (16.7 m^2/g) = 3.5 mmole/l, P = 700 mmHg, $[AlEt_3] = 50$ mmole/l, $[AlEt_2Cl] = 50$ mmole/l in 200 ml n-heptane at 41°C.)

It was supposed that structure A is formed on surface sites S_1, which are newly produced by grinding, and structure B on the intrinsic surface sites, S_2. The site A will be deactivated mainly by the attack of Cl^- from the adjacent surface sites or from the bulk of the $TiCl_3$, and to some extent by monomers or by polymer formation. The existence of sites of type A is indicated by (i) the maximum rate of polymerization per unit surface area increases with grinding work, (ii) the rate constant k_3, governing the decreasing rate of polymerization is independent of $AlEt_3$ concentration, and (iii) the activation energy of k_3, accounting for most of that of k_3, is rather low.

Structures of type B are assumed to be responsible for the observed stationary polymerization rate, and in the case of $TiCl_3/AlEt_2Cl$ probably account for the majority of the polymerization centres. A rate law can be derived as follows on the assumption that the two types of polymerization centres exist. The total number of centres [C*] at time t will be the sum of the numbers of each type, i.e.

$$[C^*]_t = [C^*_A]_t + [C^*_B] \tag{3.105}$$

The deactivation effect can be expressed as

$$-\frac{d[C^*_A]_t}{dt} = k_3[C^*_A]_t \tag{3.106}$$

with $$k_3 = k^0_3[Cl^-]_{Ti} + k^{0\prime}_3[M]$$

and from Eq. 3.105 and 3.106 we obtain

$$-\frac{d[C^*]_t}{dt} = k_3([C^*]_t - [C^*_B]) \tag{3.107}$$

Integration of Eq. 3.107 then gives

$$\frac{[C^*] - [C^*_B]}{[C^*]_0 - [C^*_B]} = e^{-k_3 t} \tag{3.108}$$

where $$[C^*]_0 = [C^*_A]_0 + [C^*_B]$$

$$= ([S_1] + [S_2])\left(\frac{K[A]}{1 + K[A]}\right) \tag{3.109}$$

These equations are in agreement with the experimental rate laws (Eq. 2.7*c*, 2.7*a* and 2.9). The activation energy of k'_3 was found to be 2 ~ 3 kcal/mole. A small activation energy of this magnitude may be expected for the migration of Cl^- from an adjacent lattice point to a surface defect such as that postulated in Cossee's model.

This mechanism will be discussed in the next chapter in terms of the isotacticity and molecular weight of the polymers obtained.

References

1. F. Eirich and H. Mark. *J. Colloid Sci.*, **11**, 748 (1956).
2. G. Natta and I. Pasquon, *Adv. Catalysis*, **11**, 1 (1959).
3. P. Cossee, *J. Catalaysis*, **3**, 80 (1964).
4. T. Keii, *Nature*, **202**, 1331 (1964); *Chem High Polymers* (Japanese), **22**, 344 (1965).
5. J. Horiuti, *J. Res. Inst. Catalysis, Hokkaido Univ.*, **1**, 1 (1948).
6. T. Keii, T. Suzuki and K. Soga, *Asahi Glass Kogyo Gizitsu Shorei Rep.* (Japanese), **12**, 155 (1966).
7. T. Keii, T. Takagi, M. Taira, K. Soga, N. Saiki and Y. Murayama, *Shokubai (Tokyo)* (Japanese), **5**, 243 (1963).
8. K. Soga, *M. Eng. Thesis, Tokyo Inst. Technology*, 1965. K. Soga, *Dr. Eng. Thesis, Tokyo Inst. Technology*, 1968. H. Koga, K. Soga and T. Keii, *Shokubai (Tokyo)* (Japanese), **7**, 2 (1965).
9. K. Fukui and T. Kagiya, presented at 12th Symposium, High Polymer Soc. Japan, 1963.
10. G. Natta, G. Mazzanti, D. De Luca, U. Giannini and F. Bandini, *Makromol. Chem.*, **83**, 54 (1964).
11. I. Pasquon, M. Dente and F. Marduzzi, *Chim. Ind. (Milan)*, **41**, 387 (1959).
12. A.D. Caunt, *J. Polymer Sci.*, **C4**, 49 (1963).
13. H. Schnecko, M. Reinmoler and K. Weirauch, *Makromol. Chem.*, **83**, 154 (1964).
14. W. E. Smith and R.G. Zelmer, *J. Polymer Sci.*, **A1**, 2587 (1963).
15. H. De Vries, *Rec. Trav. Chim.*, **80**, 866 (1961).
16. E. Kohn, H.J.L. Schuurmans, J.V. Cavender and R.M. Mendelson, *J. Polymer Sci.*, **58**, 681 (1962).
17. J. Ambroz, P. Osecky, J. Mejzlik and O. Hamerik, *J. Polymer Sci.*, **C16**, 423 (1967).
18. T. Keii, Y. Doi and I. Okura, unpublished paper.
19. T. Keii, K. Soga, S. Go, A. Takahashi and A. Kojima, *J. Polymer Sci.*, **C23**, 453 (1968).

Chapter 4

Isotacticity and molecular weight of Ziegler-Natta polymers

The stereoregularity of polymers produced with Ziegler-Natta catalysts has been the main field of interest both from a practical point of view, and in basic polymer research. It is well known that stereoregularity is a function of the nature of the catalyst, cocatalysts and additives, but in many cases quantitative data are still not available. It has often been said that the stereoregularity of a catalyst system is independent of the kinetic behavior, but some clear correlations have in fact been found.

In this chapter, experimental data on stereoregularity and molecular weight will be discussed in the light of the kinetic behavior of these catalyst systems.

4.1 STEREOREGULARITY

Stereoregularity is generally measured in two ways. The first method, due to Natta *et al.*,[1] gives the fraction (%) of polymer not extracted by boiling *n*-heptane as a measure of the stereoregularity. The second method[2] derives the I.I. (%) from the intensity ratio of the infra-red spectral bands at 995 cm^{-1} and 974 cm^{-1} for the polymer. A number of other methods exist, such as crystallinity or density measurements, but for the sake of simplicity, only the first two will be referred to in this book.

It may be noted that there is substantial disagreement between the results of the two methods in the case of polypropylene produced using $TiCl_3/ZnEt_2$ catalysts. Higher values were usually given by the

I.R. method (60–90%), whereas the boiling heptane method gave results ranging from 30–65%. It is thought[3,4] that solution of isotactic polymer of low molecular weight into the boiling heptane may be responsible for this discrepancy.

4.1.1 Stereospecificity of solid catalysts

Since the early work of Natta *et al.*,[1] it has been accepted that the α, γ and δ modifications of $TiCl_3$ all have similar specificities for α-olefins and that these specificities are the highest to be found among transition metal compounds. The β modification gives a lower specificity, comparable with that of $TiCl_4$.

From the data of Natta,[5] a number of transition metal compounds can be arranged in approximate order of decreasing specificity for propylene polymerization. The percentage of isotactic material, as found by the boiling heptane method, is shown in brackets.

$$\alpha, \gamma, \delta\text{-}TiCl_3(80\text{–}92) > VCl_3\ (73) > \beta\text{-}TiCl_3\ (40\text{–}50)$$
$$\sim TiCl_4(30\text{–}60) \sim VCl_4\ (48) > TiBr_4\ (42) > CrCl_3\ (36) \quad (4.1)$$

A similar list, based on the work of Dawans and Teyssie,[6] is given below.

$$\alpha, \gamma, \delta\text{-}TiCl_3\ (80\text{–}92) > VCl_3\ (73) > ZrCl_4\ (55) > \beta\text{-}TiCl_3\ (40\text{–}50)$$
$$\sim TiCl_4\ (48) = VCl_4\ (48) > TiBr_4\ (42) > CrCl_3\ (36) > VOCl_3\ (32) \quad (4.2)$$

This sequence is unlikely to be altered by later work, though the values obtained for isotacticity may change to some extent. Slightly different isotacticity values have been reported on the same polymerization system.

Several factors can affect isotacticity. Thus Keii *et al.*[7] found that α-$TiCl_3$ (AA-grade, prepared by Al reduction and ground) showed a slightly lower specificity than ground samples of α-$TiCl_3$ (H-grade, prepared by H_2 reduction) in a heptane solvent. This difference disappeared if toluene was used as a solvent, and was therefore attributed to the presence of $AlCl_3$ in the AA-grade $TiCl_3$. It has been shown by Boor[8] that with CdR_2, all four modifications of $TiCl_3$ exert the same, high specificity for α-olefin polymerization.

4.1.2 Influence of cocatalysts

The influence of organometallic cocatalysts on the isotacticity of polypropylene is very marked, and appears to depend on the nature of both the metal[9)] and the attached alkyl group (R).

$$AlR_3 > BeR_2 > ZnR_2 > MgR_2 > LiR \quad (4.3)$$

$$BeEt_2\ (94–96) > AlEt_3\ (80–92) > MgEt_2\ (78–85)$$
$$> ZnEt_2\ (30–40) > NaR\ (0) \quad (4.3a)$$

Recent work by Boor[8)] indicates that $CdEt_2(91) \sim ZnEt_2(63–89) > AlEt_3(77)$. The influence of the R group may be summarized from the data of Gaylord and Mark[10)] and Dawans and Teyssie.[6)]

$$Et(79.4) > i\text{-}Bu(74.5) > n\text{-}Pr(71.8) > C_6H_5(65.4)$$
$$> C_6H_{13}(64.0) > C_{16}H_{33}(59.0) \quad (4.4)$$

The substitution of a halogen atom for one of the R groups has also been shown to have a marked effect on isotacticity.[11,12)]

$$AlEt_2I(98) > AlEt_2Br(95) > AlEt_2Cl(93)$$
$$> AlEt_2F(83) \simeq AlEt_3(83) \quad (4.5)$$

$$AlEt_2I(97) > AlEt_2Br(95) > AlEt_2Cl(91)$$
$$> AlEt_2F(83) \simeq AlEt_3(83) \quad (4.5a)$$

Although $AlEt_2X$ (X = halogen) enhances stereospecificity to a very high level, its activity appears to be correspondingly reduced. According to Danusso,[11)] the relative rates of propylene polymerization by various catalyst systems, $TiCl_3/AlEt_2X$, are

$$AlEt_3(100) > AlEt_2Cl(33) \sim AlEt_2Br(33)$$
$$> AlEt_2F(30) > AlEt_2I(9) \quad (4.6)$$

The apparently reciprocal relation between specificity and activity has been quantitatively investigated by Keii *et al.*,[13)] and will be discussed in 4.2.4. Most industrial catalyst systems are based on $TiCl_3/AlEt_3$

or $TiCl_3/AlEt_2Cl$ in order to combine high specificity with a good level of activity.

The influence of additives such as amines on the stereospecificity and polymerization rate is also important. This will be discussed in the next chapter.

4.2 STEREOSPECIFICITY AND KINETIC BEHAVIOR OF TWO-COMPONENT ZIEGLER-NATTA CATALYST SYSTEMS

Contrary to earlier work,[1] recent experimental data have indicated some correlation of stereospecificity with polymerization conditions (i.e. kinetic behavior). The experimental results of Natta *et al.*[1] (using α-$TiCl_3$ which was ground in a stainless steel ball mill to a particle size of $\leq 2\mu$ and washed several times with *n*-heptane) are shown in Table 4.1. From these results, Natta *et al.* concluded that the polymerization time did not influence the molecular weight or steric composition of the polymer. Other polymerization conditions were also examined. At concentrations of $AlEt_3$ ranging from 11.8–147.5 mmole/l

TABLE 4.1 Stereoregularity and molecular weight under various propylene polymerization conditions ($TiCl_3$–$AlEt_3$)[1]

T (°C)	P (mm Hg)	[Al] (mole/l)	α-$TiCl_3$ (sample A) (g/l)	[A]/[Ti]	Polymerization time (hr)	Non-atactic polymer (%)†	$[\eta]$ (100 cm^3/g) for non-atactic polymer
31	700	2.94×10^{-2}	1.50	3	14	89	4.40
					31	88.5	4.34
31	1450	2.94×10^{-2}	3.00	1.5	1	85	4.66
					4	86	4.65
31	1450	2.94×10^{-2}	1.50	3	8	86	4.70
					17	84	4.90
51	1110	1.47×10^{-2}	0.30	7.5	8	87	4.50
					24	87	4.45
70	450	2.36×10^{-2}	1.20	3	10	90	3.56
					15	90	3.47
70	450	2.94×10^{-2}	1.50	3	4	91	3.28
					7	91	3.14

TABLE 4.1—*Continued*

T (°C)	P (mm Hg)	[Al] (mole/1)	α-$TiCl_3$ (sample A) (g/l)	[A]/[Ti]	Polymerization time (hr)	Non-atactic polymer (%)†	[η] (100 cm^3/g) for non-atactic polymer
70	450	7.36×10^{-2}	11.30	1	1/3	91	2.18
					1	90	2.13
70	950	2.94×10^{-2}	1.50	3	1/2	88.5	3.84
					2	88.5	3.78
					6	88	3.84
70	1450	1.77×10^{-2}	0.91	3	4	88	4.16
					7	87	4.20

† Polymer insoluble in *n*-heptane at room temperature. The data then include stereoblock polymers soluble in boiling *n*-heptane (5–7%).

(1 < [Al]/[Ti] < 10) and pressures of 450–1450 mmHg at various temperatures, it was found that the fraction of non-atactic polymer remained almost constant at 88–90%. The only exception was that data obtained at 1110 mm Hg and 51 °C showed an increase from 83% to 89% with increasing $AlEt_3$ concentration (14.7 → 117.5 mmole/l), though this tendency was not noted by Natta *et al.*

Natta and Pasquon did, however, confirm that increases in that fraction of polymer soluble in boiling ether or *n*-heptane took place with decreasing propylene pressure (1600 → 250 mm Hg) and increasing $AlEt_3$ concentration (15 → 120 mmole/l). This was considered to be due to the inversion of configuration in the monomer units which occurred during the growth of atactic or block polymeric chains. Changes in the rate of inversion due to differing propylene pressures or $AlEt_3$ concentrations were attributed to alteration in the relative rates of the transfer and propagation reactions.

4.2.1 Effect of organometal concentration on stereospecificity

Boor[8] noted a clear tendency for specificity to increase with decreasing concentration of $ZnEt_2$, as shown in Table 4.2, and a similar tendency was found by Ingberman *et al.*[14] in the case of $AlEt_3$ (Table 4.3). More detailed experiments have been carried out by Keii, Soga and coworkers,[7] who investigated the influence of the solvent, temperature and $AlEt_3$ concentration on polypropylene isotacticity.

TABLE 4.2 Polymerization of propylene and butene-1[8)†1]

	$[ZnEt_2]$ (mmole)	Conversion (ωt %)	isotacticity (%)
Propylene			
	1.0 (α-$TiCl_3$)		89
	6.3 (β)	45	79
	1.0 (β)	20	68
	5.0 (γ)	80	68
	1.0 (γ)	20	82
	5.0 (γ)	80	63
Butene-1			
	1.5 (γ)	8	92†2
	2.5 (γ)	13	77†2
	6.0 (γ)	48	57†2

†1 8 oz. bottle, 1.5 mmole-$TiCl_3$, 50°C, 100 ml, 20 hr.
†2 Percent insoluble in boiling ether.

TABLE 4.3 Effect of aluminium alkyl and concentration on polymer yield and tacticity[14)†1]

Alkyl	Conc. (mmole/l)	Polymer yield (g/mmole-Ti)	Amorphous (%)	Crystallinity†2 (%)	$\bar{M}_v \times 10^{-5}$
$AlEt_2Cl$	0.256	3.61	1.5	85	12.5
	0.565	4.75	0	85	14.0
$AlEt_3$	0.260	3.02	6.1	85	17.2
	0.442	8.88	7.0	70	13.6
	0.700	13.0	13.2	65	

†1 40°C, 75 psig propylene, 2.5 hr, 2.0 ± 0.1 mmole $TiCl_3$, *n*-heptane.
†2 Refers to the crystallinity of polymer after removal of the wholly amorphous fraction based on the infrared method of Luongo.[2)]

Fig. 4.1 shows the dependence of I.I. (%) on $AlEt_3$ concentration at various temperatures. Open circles indicate data obtained on the $(AlEt_3 + SeOCl_2)/TiCl_3$ system, using corrected concentrations.

4.2.2 Effect of temperature on stereospecificity

The results of Keii *et al.*,[7)] are given in Fig. 4.2. The specificity of the system (ground AA-grade $TiCl_3/AlEt_3$) increases with temperature

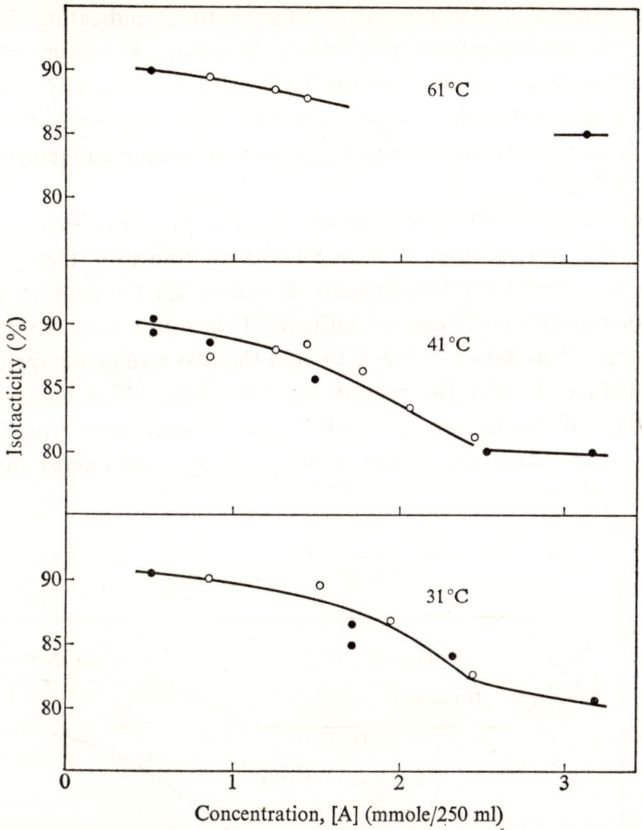

Fig. 4.1. Relationship between isotacticity and effective concentration of $AlEt_3$ for propylene polymers. (3.2 mmole $TiCl_3$, 3.2 mmole $AlEt_3$ after 2 hr in 250 ml toluene, in the presence (○) and absence (●) of $SeOCl_2$).

increase above 55 °C and increases with temperature decrease below 30 °C, whereas that of $TiCl_3/AlEt_2Cl$ remains unchanged. Furthermore it was found that the specificity of AA-grade $TiCl_3$ is higher in toluene than in *n*-heptane. Such a solvent effect was not found for H-grade $TiCl_3$.

In connection with the temperature effect on stereospecificity, the kinetic curves of the same systems were examined, and the results are shown in Fig. 4.3. Curves A and B were obtained at 67 °C and 43 °C respectively. On changing the temperature from 43 °C to 67 °C, the kinetic curve was displaced from B to A, but the reverse temperature change resulted in a displacement from A to C, indicating that the change in polymerization rate was irreversible. However, changes occurring with the catalyst system $TiCl_3/AlEt_2Cl$ were fully reversible as can be seen in Fig. 4.3 (Curves D and E). This behavior is illustrated by the relation between stationary polymerization rate and temperature shown in Fig. 4.4.

The temperature coefficient obtained with $AlEt_3$ below about 50°C is the same as that of $AlEt_2Cl$ and an activation energy of 10 kcal/mole is found for both systems. However, the temperature coefficient for the stationary rate of $AlEt_3/TiCl_3$ has also been reported by Natta *et al.*,[1] as shown in Fig. 2.19, and the two results are apparently contradictory. In fact the catalyst used by Keii differs considerably from that of Natta, whose catalyst gave acceleration type kinetic curves, rather than the higher activity, decay type curves of Keii.

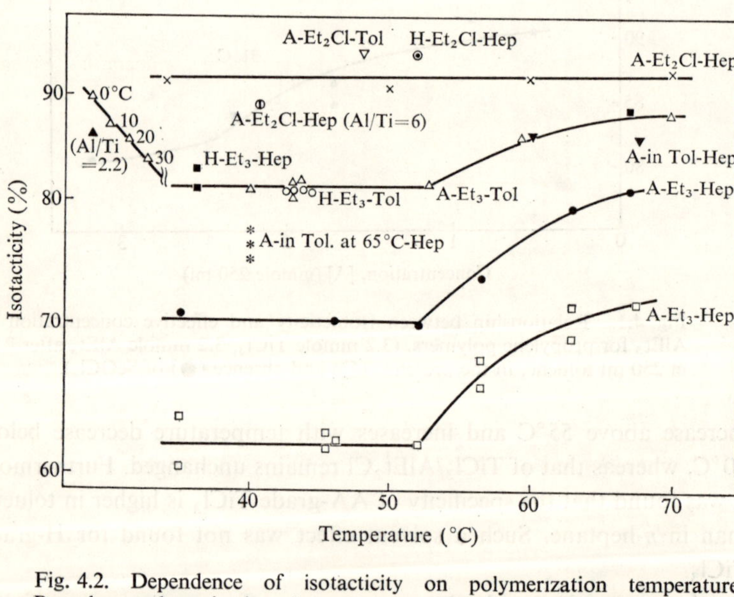

Fig. 4.2. Dependence of isotacticity on polymerization temperature. Propylene polymerization time, 90 min, [A]/[Ti] = 1. (Stauffer AA-$TiCl_3$/$AlEt_2Cl$/*n*-heptane, ×; AA-$TiCl_3$/toluene, △; HA-$TiCl_3$/$AlEt_3$/toluene, ○; HA-$TiCl_3$/*n*-heptane, ■; AA-$TiCl_3$/$AlEt_3$, *n*-heptane, toluene, ▼; AA-$TiCl_3$ (treated with toluene at 65°C)/$AlEt_3$/*n*-heptane, *; AA-$TiCl_3$/$AlEt_3$/*n*-heptane, ●; polymer unextractable in boiling *n*-heptane, AA-$TiCl_3$/$AlEt_3$, □; AA-$TiCl_3$/$AlEt_2Cl$/*n*-heptane ([A]/[Ti] = 6), ⦶; AA-$TiCl_3$/$AlEt_2Cl$/toluene, ▽; HA-$TiCl_3$/$AlEt_2Cl$/*n*-heptane, ⊙.)

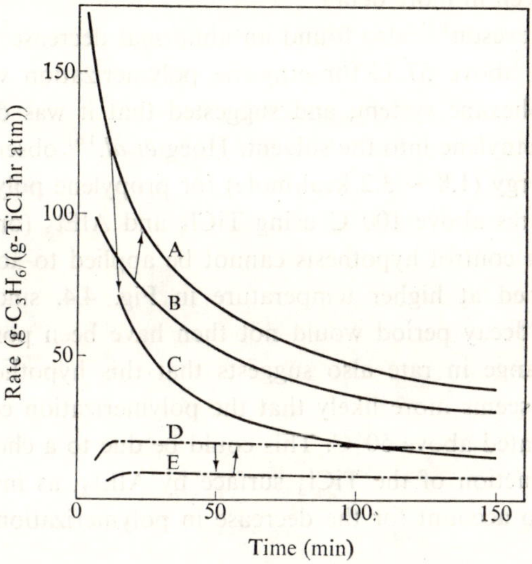

Fig. 4.3. Kinetic curve transitions produced by temperature changes. ([$TiCl_3$] = 2 g/l, [A] = 15 mmole/l in 250 ml toluene. A, 67°C and B, C, 43°C for $AlEt_3$; D, 67°C and E, 43°C for $AlEt_2Cl$.)

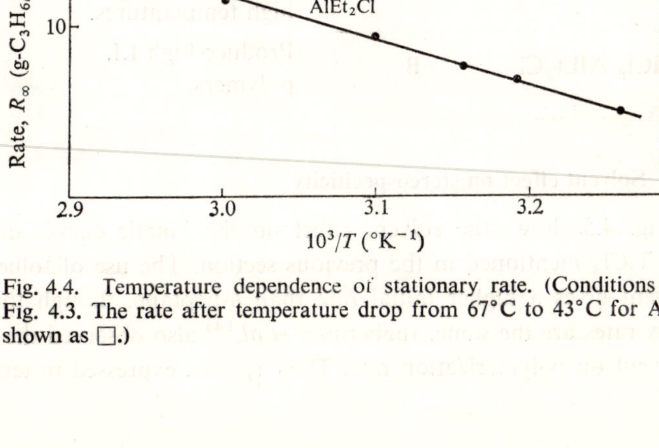

Fig. 4.4. Temperature dependence of stationary rate. (Conditions as for Fig. 4.3. The rate after temperature drop from 67°C to 43°C for $AlEt_3$ is shown as □.)

It may therefore be worthwhile to discuss the abnormal temperature effect observed by Keii in more detail.

Berger and Grieveson[15)] also found an abnormal decrease in the polymerization rate above 57°C for ethylene polymerization with a $TiCl_3/AlEt_2Cl$/cyclohexane system, and suggested that it was due to slow absorption of ethylene into the solvent. Hoeg *et al.*[16)] observed a small activation energy (1.8 ~ 3.2 kcal/mole) for propylene polymerization at temperatures above 100°C using $TiCl_3$ and $AlEt_3$ (or Al(*i*-Bu)$_3$). The diffusion control hypothesis cannot be applied to account for the data obtained at higher temperature in Fig. 4.4, since the higher rates of the decay period would not then have been possible. The irreversible change in rate also suggests that this hypothesis is inadequate here. It seems more likely that the polymerization centres are partially deactivated above 50°C. This could be due to a chemical process such as reduction of the $TiCl_3$ surface by $AlEt_3$, as invoked by Natta *et al.*[17)] to account for the decrease in polymerization with $VCl_3/AlEt_3$.

In view of the temperature dependence of polymerization rate and isotacticity, Keii *et al.*[18)] proposed the existence of two types of polymerization centre, along the following lines.

Catalyst system	Polymerization centres	Properties of polymerization centres
$TiCl_3/AlEt_3$	A	A; unstable at high temperature. Produce low I.I. polymers.
	B	B; stable even at high temperatures.
$TiCl_3/AlEt_2Cl$	B	Produce high I.I. polymers.

4.2.3 Solvent effect on stereospecificity

Fig. 4.5 shows the solvent effect on the kinetic curves of AA-grade $TiCl_3$ mentioned in the previous section. The use of toluene as a solvent gives a higher initial rate than *n*-heptane, though the stationary rates are the same. Ingberman *et al.*[14)] also observed the effect of solvent on polymerization rate. Their results, expressed in terms of

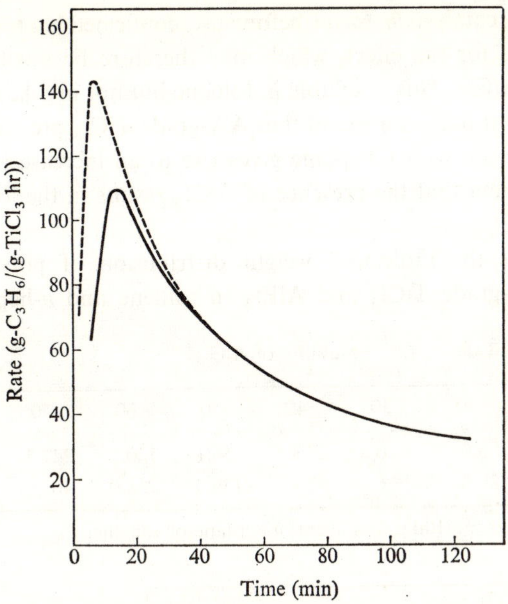

Fig. 4.5. Solvent effect on kinetic curves (·······shows toluene and ——is *n*-heptane.)

TABLE 4.4 Effect of solvent on polymerization rate†

Solvent	Dielectric constant	*n*	k''(g/mmole-$TiCl_3$)
n-Heptane	1.9	0.41	2.5
Toluene	2.3	0.19	4.7
Chlorobenzene	5.4	0.27	4.6
o-Dichlorobenzene	9.3	0.28	4.8

† 40°C, 7.5 psig propylene, 2.5 hr, $AlEt_2Cl$, 1.5 l solvent. The α/γ $TiCl_3$ used was obtained by heating β-$TiCl_3$ prepared by reduction of $TiCl_4$ with Al(i-Bu)$_3$.

polymer yield, $R = k''[AlEt_2Cl]^n$, are shown in Table 4.4. Shilov *et al.*[19)] investigated ethylene polymerization with the homogeneous system $(C_5H_5)_2TiCl_2/AlMe_2Cl$, and found that with toluene the initial rate build-up was faster than with *n*-heptane. They accounted for this on the basis of the different dielectric constants of the two solvents.

The solvent effect observed by Keii *et al.* with AA-grade $TiCl_3$ was independent of the nature of the cocatalyst. Since AA-grade $TiCl_3$ is prepared by the reduction of $TiCl_4$ with Al, it contains 25 mole%

$AlCl_3$. Heating the catalyst *in vacuo* before use confirmed that $TiCl_4$ was not responsible for the effect, which may therefore be attributed to the presence of $AlCl_3$. This is soluble in toluene but not in *n*-heptane (see Table 4.5), and it has been found that AA-grade $TiCl_3$ pre-treated with toluene before use with *n*-heptane gives rise to an increase in I.I. It is therefore apparent that the presence of $AlCl_3$ results in the loss of some specificity.

Fig. 4.6 shows the molecular weight distribution of polymers produced with AA-grade $TiCl_3$ and $AlEt_3$ in toluene and *n*-heptane

TABLE 4.5 Solubility of $AlCl_3$†

Temp. (°C)	20	30	40	50	60	70
n-Heptane	5.0	12.0	27.3	56.1	120.3	227.7
Toluene	39.0	48.6	82.7	139.9	258.0	407.6

† Expressed as mg/100 g of solvent, *n*-heptane or toluene.

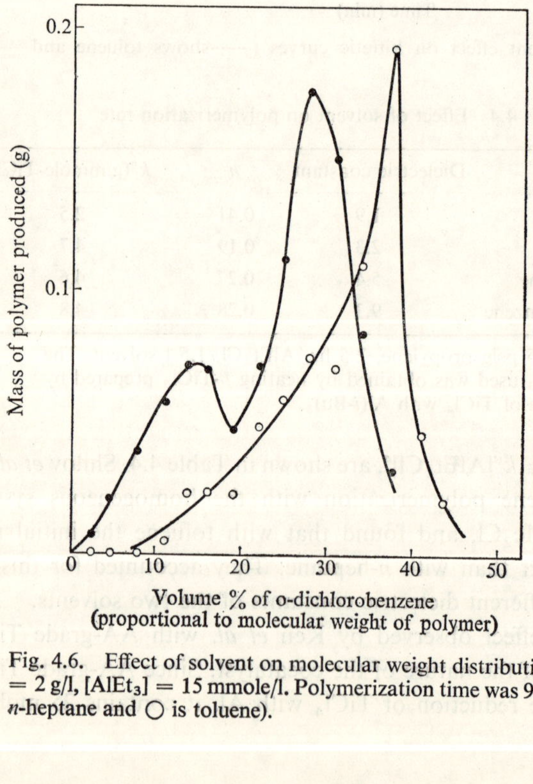

Fig. 4.6. Effect of solvent on molecular weight distribution at 44°C. ($[TiCl_3]$ = 2 g/l, $[AlEt_3]$ = 15 mmole/l. Polymerization time was 90 min; ● represents *n*-heptane and ○ is toluene).

solvents. In the case of *n*-heptane, 30% of the polymer was of such low molecular weight that the I.I. values could not be examined. Again, this appears to be due to the presence of $AlCl_3$.

4.2.4 Ageing effect on stereospecificity

The decay type kinetic curves obtained with active $TiCl_3/AlEt_3$ catalyst systems are strongly affected by the ageing of the $TiCl_3$ in a concentrated solution of $AlEt_3$ where $[Al]/[Ti] \geq 10$ (see (3.4), Fig. 3.4 and 3.5). Keii *et al.*[13)] found that in addition to the decrease in maximum polymerization rate, the I.I. of the polymer increased with increasing ageing time. This ageing effect can therefore be regarded as due to deactivation of some of the polymerization centres. Other factors which act to decrease the maximum rate of polymerization, such as addition of Et_3N to the catalyst system or long-term storage of the $TiCl_3$, usually also cause some increase in stereospecificity. The relationship may be expressed as follows.

$$\text{I.I.} \propto \frac{1}{R} \tag{4.7}$$

An equation of this form can be interpreted in terms of the present author's assumption of two types of polymerization centre (4.2.2). The decay period of the kinetic curve can be expressed by Eq. 3.108 or 2.7*a*.

$$\frac{[\mathrm{C}^*] - [\mathrm{C}^*_\mathrm{B}]}{[\mathrm{C}^*]_0 - [\mathrm{C}^*_\mathrm{B}]} = e^{-k_3 t} \tag{3.108}$$

or

$$R = k_p P[([\mathrm{C}^*]_0 - [\mathrm{C}^*_\mathrm{B}])e^{-k_3 t} + [\mathrm{C}^*_\mathrm{B}]]$$

$$= R_\infty + (\bar{R}_0 - R_\infty)e^{-k_3 t}$$

If I_L represents the I.I. value of polymers produced on the A polymerization centres and I_H that of polymer produced on the B centres, then the average value $I(t)$ of the I.I. of polymers produced between the times t and $t + dt$ can be expressed as

$$I(t) = \frac{I_H R_\infty + I_L(R - R_\infty)}{R}$$

$$= I_L + \frac{(I_H - I_L)R_\infty}{R} \tag{4.7a}$$

The value of I.I. usually measured is the integral value, I.I. (t), over the total polymerization time

$$\text{I.I.}(t) = \frac{\int_0^t I(t)R\,dt}{\int_0^t R\,dt}$$

$$= I_L + \frac{(I_H - I_L)R_\infty}{t^{-1}\int_0^t R\,dt} \tag{4.7b}$$

where $t^{-1}\int_0^t R\,dt$ is the polymer yield per unit time (hr). Both Eq. 4.7*a* and 4.7*b* are compatible with the reciprocity relation (Eq. 4.7). Keii *et al.*[13]) measured the isotacticity of polymers produced with $TiCl_3$ aged in $AlEt_3$, with old samples of $TiCl_3$, and with catalyst systems containing Et_3N, and obtained the results shown in Fig. 4.7.

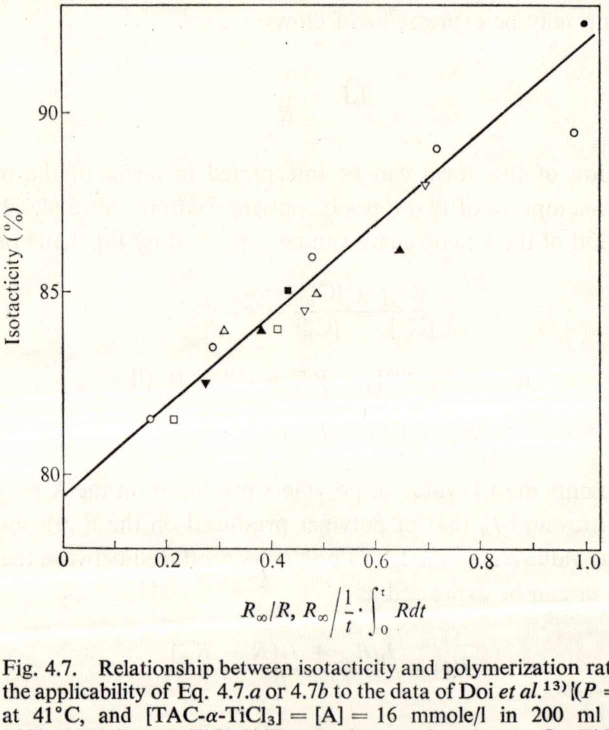

Fig. 4.7. Relationship between isotacticity and polymerization rate, showing the applicability of Eq. 4.7.*a* or 4.7*b* to the data of Doi *et al.*[13]) (P = 70 cmHg at 41°C, and [TAC-α-$TiCl_3$] = [A] = 16 mmole/l in 200 ml *n*-heptane. $TiCl_3/AlEt_2Cl$, ●; $TiCl_3/AlEt_3$ (various ageing times), ○; $TiCl_3$ (treated with conc. $AlEt_3$)/$AlEt_3$, ▲; $TiCl_3$ (treated with conc. $AlEt_2Cl$)/$AlEt_3$, ▼; $TiCl_3$ + $AlEt_3$ (integral), □; $TiCl_3$ (stored for 2 years)/$AlEt_3$, △; $TiCl_3$ (treated with Et_3N)/$AlEt_3$, ■; $TiCl_3/AlEt_3$, gas-phase polymerization, ▽.)

TABLE 4.6 Influence of amines on the solution polymerization of propylene†1

Molar ratios (Ti: Al: amine)	Amine	Polymerization rate (g-C_3H_6/(g-$TiCl_3$ hr))	Polymer unextractable in boiling *n*-heptane (%)
1:1:0	0	154	67
1:1:1	$(C_4H_9)_2NH$	44	86
1:1:1	$(C_2H_5)_3N$	122	75
1:1:3	$(C_2H_5)_3N$	51	85
1:1:5	$(n\text{-}C_4H_9)_3N$	117	77
1:1:20	$(n\text{-}C_4H_9)_3N$	6	94
1:1:1†2	Pyridine	45	72
1:1:1	2,4,6-Trimethyl-pyridine	117	75
1:1:5†3	$(n\text{-}C_4H_9)_3N$	150	81
1:1:20†3	$(n\text{-}C_4H_9)_3N$	30	94

†1 120~130°C, $P_0 = 9.45$ kg/cm², 385 ml cyclohexane, 0.43 g $TiCl_3$. At 120°C, 0.38 g of $AlEt_3$ in cyclohexane was added.
†2 Use of pyridine was restricted to *ca.* 86°C.
†3 $AlEt_3$ used; $AlEt_3 \cdot$ amine added after start of polymerization.

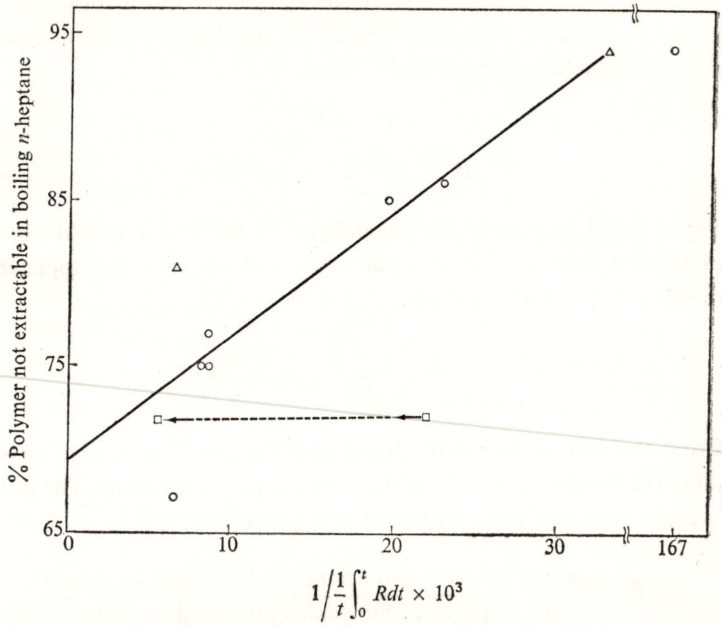

Fig. 4.8. Relationship between isotacticity and polymerization rate from the data of Hoeg and Liebman.[20] (see Table 4.6)

Thus Eq. 4.7 can be substantiated, at least for the $TiCl_3/AlEt_3$ catalyst system. The experimental data of Hoeg and Liebman[20] are summarized in Table 4.6, showing the effects of various amines on polymer isotacticity. Fig. 4.8 shows this data in terms of Eq. 4.7*b*, with the data at 86 °C corrected to 120–130 °C (where the activation energy is 10 kcal/mole). This also provides some support for Eq. 4.7 and for the two centre polymerization theory. The increase in specificity which accompanies a decrease in $AlEt_3$ concentration or polymerization temperature may also be governed by a reciprocal relationship.

4.3 MOLECULAR WEIGHT AND POLYMERIZATION CONDITIONS

4.3.1 Molecular weight during the adjustment period

The molecular weight of polymers produced with α-$TiCl_3$ and $AlEt_3$ increases gradually with polymerization time until a stationary value is reached. The first observations of this type of change in isotactic polymers were made by Natta *et al.*[1] Their results have been summarized on the basis of a relationship between the intrinsic viscosity, η, and viscometric molecular weight, $\overline{M}_v$ (or degree of polymerization, $\bar{P}_n$), for isotactic polypropylene

$$[\eta] = K\overline{M}_v^{0.74} \tag{4.8}$$

$$\bar{P}_n \propto [\eta]^{1/0.74} = [\eta]^{1.35} \tag{4.8a}$$

Fig. 4.9 and 4.10 show the dependence of molecular weight on polymerization time for isotactic polymers, and illustrate the applicability of the experimental equation

$$\frac{1}{[\eta]^{1.35}} = \frac{1}{[\eta]_\infty^{1.35}} + \frac{\text{const.}}{t} \tag{4.9}$$

From the data in Fig. 4.10, Natta and Pasquon[1] were able to determine the mean lifetime (L) of the growing polymer chains. The number average degree of polymerization at a given time, t, is

$$\bar{P}_n = \frac{\text{no. of monomer molecules polymerized by time } t}{\text{no. of polymer molecules produced by time } t}$$

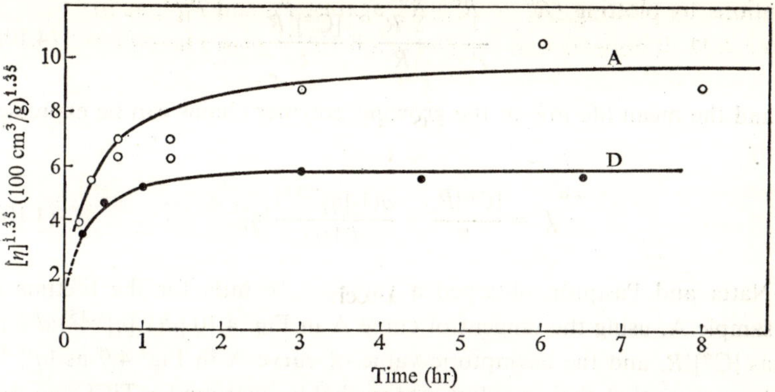

Fig. 4.9. Plot of intrinsic viscosity of the non-atactic polypropylene fraction to the power 1.35 *vs.* polymerization time.[1] ([$AlEt_3$] = 23.5 mmole/l, two samples of ground α-$TiCl_3$ at 15°C and 200 mmHg.)

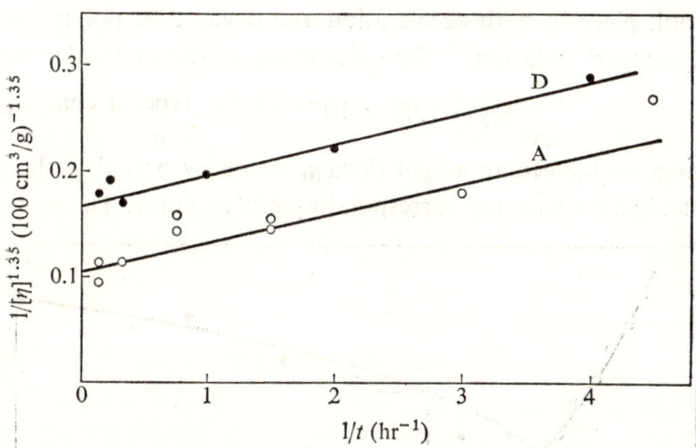

Fig. 4.10. Relationship between the reciprocals of intrinsic viscosity to the power 1.35, and polymerization time.[1]

$$= \frac{\int_0^t R dt}{[C^*]_t + \int_0^t \Sigma R_t dt} \tag{4.10}$$

where R is the rate of polymerization, ΣR_t is the sum of the rates of chain transfer and termination processes, and $[C^*]_t$ is the number of growing polymer chains at time t. If R, R_t and $[C^*]_t$ are all independent of time, then

$$\frac{1}{\bar{P}_n} = \frac{\Sigma R_t}{R} + \frac{[C^*]/R}{t} \tag{4.10a}$$

and the mean lifetime of the growing polymer chains can be expressed by

$$L = \frac{[C^*]\bar{P}_n}{R} = \frac{d(1/[\eta]^{1.35})}{d(1/t)}[\eta]_\infty^{1.35} \tag{4.10b}$$

Natta and Pasquon obtained a value of 16 min. for the lifetime in sample A, using the tangent of curve A in Fig. 4.10 $[d(1/[\eta]^{1.35})/d(1/t)]$ as $[C^*]/R$, and the asymptotic value of curve A in Fig. 4.9 as $[\eta]^{1.35}$. It appears that their catalyst was probably unground α-$TiCl_3$, so that the kinetic curve was of the acceleration type during the adjustment period.

Murayama and Keii[21] examined the change in molecular weight which took place in both acceleration and decay type polymerization using Kinsinger's relation[22] for calculation of the molecular weight.

$$[\eta] = 1.07 \times 10^{-4}\ \bar{M}_v^{0.80} \tag{4.11}$$

The change in molecular weight during the decay period is shown in Fig. 4.11 together with the corresponding kinetic curve. The molecular

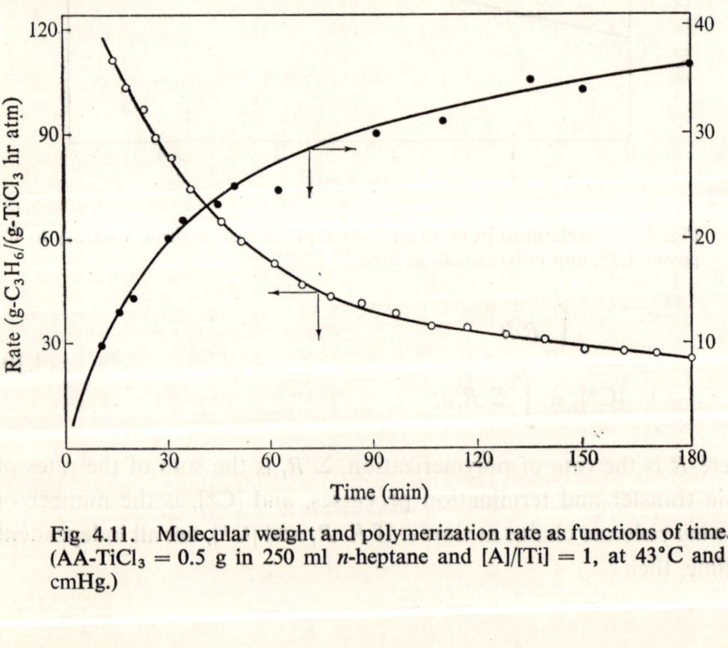

Fig. 4.11. Molecular weight and polymerization rate as functions of time.[21] (AA-$TiCl_3$ = 0.5 g in 250 ml *n*-heptane and [A]/[Ti] = 1, at 43°C and 70 cmHg.)

weight changes were also examined for those fractions of total polymer extractable in hot ether and hot *n*-heptane, and for the fraction which was insoluble in *n*-heptane. The results are shown in Fig. 4.12, 4.13 and 4.14 respectively.

The mean lifetime of the polymer chains is shown in Fig. 4.15,

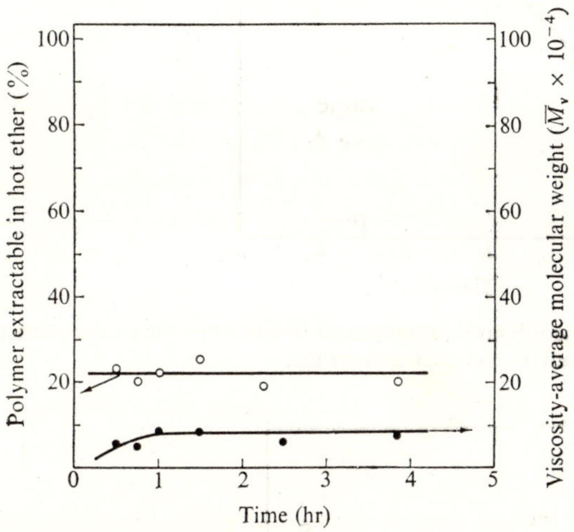

Fig. 4.12. Viscosity-average molecular weight and polymer extractable in hot ether (%) *vs.* polymerization time.[21)]

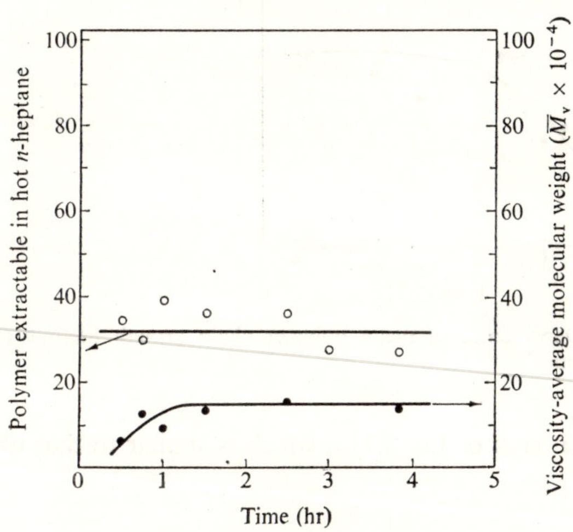

Fig. 4.13. Viscosity-average molecular weight and polymer extractable in boiling *n*-heptane (%) *vs.* polymerization time.[21)]

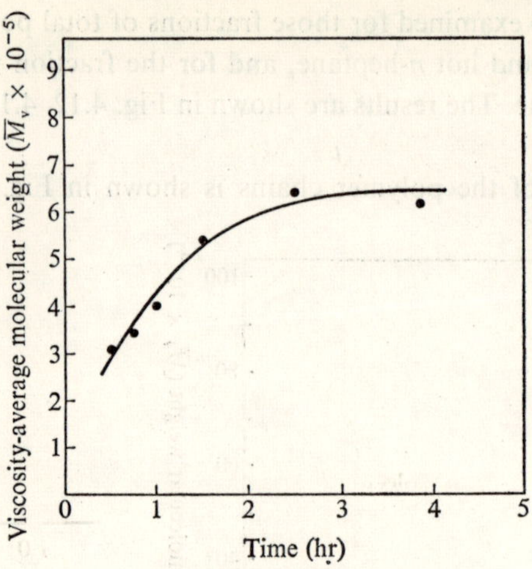

Fig. 4.14. Viscosity-average molecular weight of polymers not extractable in boiling *n*-heptane *vs.* polymerization time.[21]

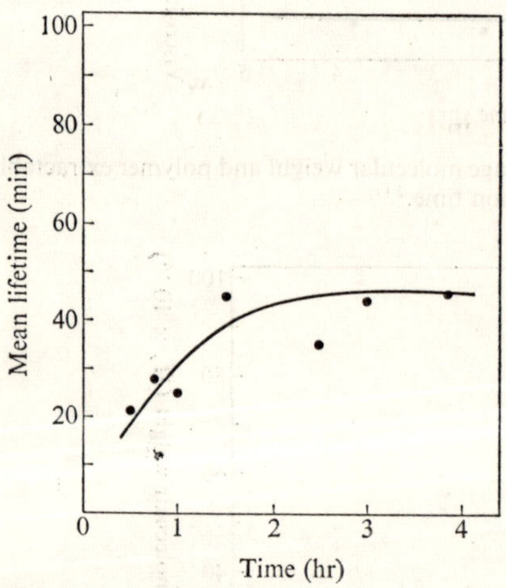

Fig. 4.15. Dependence of mean polymer chain growth lifetime on polymerization time.[21]

calculated on the basis of Eq. 4.11*a*, which is similar to that of Natta and Pasquon.[24]

$$L = \frac{d(1/[\eta]^{1.20})}{d(1/t)} [\eta]_{\infty}^{1.20} \tag{4.11a}$$

The dependence of molecular weight on polymerization time, as shown in Fig. 4.11, can be represented by

$$\frac{1}{\overline{M}_v} = \frac{1}{M_\infty} + \frac{\beta}{t} \tag{4.12}$$

$$\overline{M}_v = M_\infty(1 - e^{-\alpha t}) \tag{4.12a}$$

where α and β are constants. Table 4.7 shows the data obtained with AA grade $TiCl_3$ and $AlEt_3$.

TABLE 4.7 Viscosity-average molecular weight $\overline{M}_v$ and values of α†

A/Ti	0.2	0.3	0.4	0.6	1.0
$\overline{M}_v \times 10^{-5}$	7.5	7.0	5.0	4.2	3.5
$\alpha(\text{hr}^{-1})$	0.97	0.84	(1.6?)	1.15	1.04

† 44°C, 65 cmHg, 2 g-$TiCl_3$/l *n*-heptane.

The stationary values of molecular weight shown in Table 4.7 are compatible with the relation found by Natta *et al.*[1)]

$$\frac{1}{M_\infty} \propto [\text{A}]^{1/2} \tag{4.13}$$

Stationary state molecular weights will be discussed in the following section.

During the initial stages of polymerization, the rate increases with time such that $R_0 \propto t$, and this relation may be regarded as correct until the maximum rate is reached. In the case of decay type polymerization, we will have

$$R = R_m\left(\frac{t}{t_m}\right) \qquad t \leq t_m \tag{4.14}$$

where t_m is the time at which the rate reaches the maximum value R_m. The molecular weight can be approximated as

$$\overline{M}_v = M_\infty \alpha t \tag{4.15}$$

according to Eq. 4.12*a* for $t \leq t_m$. From Eq. 4.14 and 4.15, the number of polymer chains, N, produced by time t will be approximately

$$N = \frac{\int_0^t R\,dt}{M_v}$$

$$\simeq \frac{R_m(t^2/2t_m)}{M_\infty \alpha t}$$

$$= \left(\frac{R_m}{2\alpha M_\infty}\right)\frac{t}{t_m} \qquad (4.16)$$

In the case of propylene polymerization with AA-grade $TiCl_3$ and $AlEt_3$ ([A]/[Ti] = 1) at 44°C and 66 cm Hg, the experimental data are; $R_m = 110$ g-C_3H_6/(g-$TiCl_3$ hr atm) at $t_m = 15$ min, with $M_\infty = 3.5 \times 10^5$ and $\alpha = 1.0$ (Table 4.5). Applying Eq. 4.16, at time $t = t_m$, we obtain

$$N_m = \frac{R_m}{2\alpha M_\infty}$$

$$\simeq 10^{20}\ (1/\text{g-TiCl}_3)$$

The specific surface area of the $TiCl_3$ used was found to be 20 m^2/g by the BET method, and the maximum number of growing chains per unit surface area can therefore be obtained as

$$N_m \simeq 5 \times 10^{14}\ \text{cm}^{-2} \qquad (4.17)$$

which is comparable with the total number of Ti atoms present per unit surface area.

The polymerization rate decreases from the maximum to the stationary value according to the following equation (Eq. 2.7*a*).

$$R = R_\infty + (\bar{R}_0 - R_\infty)e^{-k_3 t} \qquad (4.18)$$

With the integral form of this expression and either Eq. 4.12*a* or the experimental molecular weight, we may calculate values of N at $t \geq t_m$.

The change in the number of growing chains during polymerization may also be expressed in the following manner. From the first equation for N in Eq. 4.16 we have

$$\frac{dN}{dt} = \frac{R\overline{M}_v - \left(\frac{d\overline{M}_v}{dt}\right)\int_0^t R\,dt}{M_1^2} = \frac{d[C^*]}{dt} + R_t \qquad (4.19)$$

The values of dN/dt calculated from the experimental data form a curve which is concave upwards and has a minimum, as illustrated in Fig. 4.16. The section A-B may be considered to correspond to the increase,

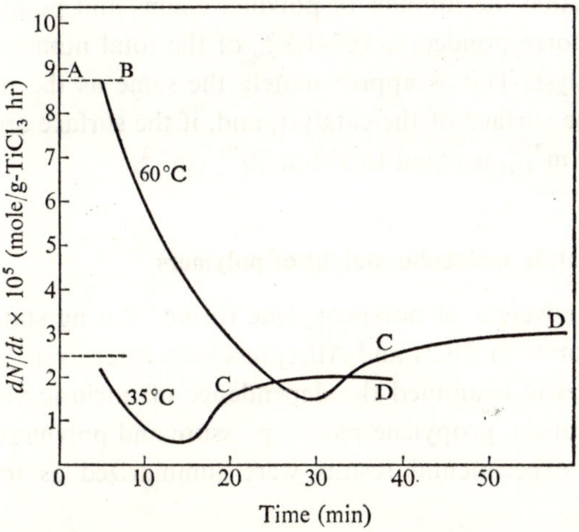

Fig. 4.16. Relationship between dN/dt and polymerization time from Eq. 4.19.[21)]

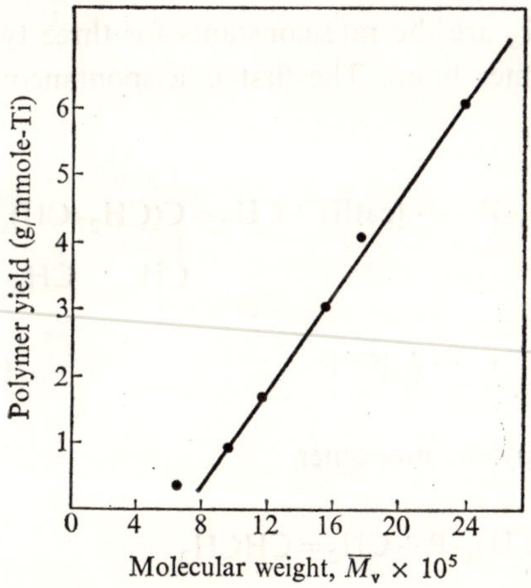

Fig. 4.17. Viscosity-average molecular weight *vs.* polymer yield.[23)]

$d[C^*]/dt$, in the number of polymerization centres, and the flat region C-D to the transfer reaction rate R_t (in numbers of polymer chains). The results suggest that the transfer reaction starts in the later stages of polymerization.

Ingberman, Levine and Turbett[14)] indicated that no transfer occurred with $TiCl_3$ and $AlEt_2Cl$ since the polymer yield was linearly dependent on molecular weight (Fig. 4.17). Assuming that $M_v/M_n = 2 \sim 5$, they calculated the number of polymer chains and pointed out that the number corresponded to 0.7–1.8% of the total number of Ti atoms in the catalyst. This is approximately the same as the number of Ti atoms on the surface of the catalyst, and, if the surface area was of the order of 10 m^2/g, is equal to about 10^{14} cm^{-2}.

4.3.2 Stationary state molecular weight of polymers

The molecular weight of polypropylene formed during stationary state polymerization with $TiCl_3$ and $AlEt_3$ has been extensively studied by Natta *et al.*,[1)] who examined the dependence of intrinsic viscosity on $AlEt_3$ concentration, propylene partial pressure and polymerization temperature. The experimental results were summarized as follows

$$\frac{1}{M_\infty} = \frac{k_1 + k_2P + k_3[A]^{1/2}}{k'_pP} \tag{4.20}$$

The constants k_1, k_2 and k_3 are the rate constants for three types of fission of the catalyst-polymer bond. The first is a spontaneous dissociation

$$[cat]\,CH_2\text{–}\underset{\textstyle CH_3}{\underset{|}{CH}}(CH_2\text{–}\underset{\textstyle CH_3}{\underset{|}{CH}})_n\text{–}R \longrightarrow [cat]H + CH_2{=}\underset{\textstyle CH_3}{\underset{|}{C}}(CH_2\text{–}\underset{\textstyle CH_3}{\underset{|}{CH}})_n\text{–}R$$

where

$$r_1 = k_1[C^*] \tag{4.20a}$$

the second is fission by propylene monomer

$$[cat]\,CH_2\text{–}\underset{\textstyle CH_3}{\underset{|}{CH}}(CH_2\text{–}\underset{\textstyle CH_3}{\underset{|}{CH}})_n\text{–}R + CH_2{=}CHCH_3$$

$$\longrightarrow \text{[cat] } CH_2CH_2CH_3 + CH_2{=}\underset{\displaystyle CH_3}{C}(CH_2{-}\underset{\displaystyle CH_3}{CH})_n{-}R$$

where
$$r_2 = k_2P[C^*] \tag{4.20b}$$

and the third is due to $AlEt_3$ monomer

$$\text{[cat] } CH_2{-}\underset{\displaystyle CH_3}{CH}(CH_2{-}\underset{\displaystyle CH_3}{CH})_n{-}R + AlEt_3$$

$$\longrightarrow \text{[cat]}CH_2CH_3 + Et_2Al{\cdot}CH_2\underset{\displaystyle CH_3}{CH}(CH_2{-}\underset{\displaystyle CH_3}{CH})_n{-}R$$

where
$$r_3 = k_3[C^*][A]^{1/2} \tag{4.20c}$$

The term $[A]^{1/2}$ in Eq. 4.20*c* arises because the concentration of $AlEt_3$ monomer is represented by

$$[AlEt_3] = (K[Al_2Et_6])^{1/2} = \left(\frac{K[A]}{2}\right)^{1/2} \tag{4.20d}$$

The existence of the process shown in Eq. 4.20*c* has been questioned by Hoeg and Liebman,[16] who were unable to detect any chain transfer processes involving organo-aluminium compounds during propylene polymerization. Instead, Hoeg[23] considered the possibility that some soluble species might be present in $TiCl_3$ that would be active either alone or combined with $AlEt_3$. However, in the case of polymerization with $TiCl_3$ and $ZnEt_2$, the role of $ZnEt_2$ as a chain-transfer agent is clear. The observed effects of $ZnEt_2$ concentration are shown in Fig. 4.18.[26] The relationships are linear as might be expected, since $ZnEt_2$ is monomeric in solution. The slope of the lines suggests that $ZnEt_2$ is a strong chain transfer reagent, even though it can readily be removed from the catalyst system by washing with solvent, indicating that its affinity for the $TiCl_3$ surface is weaker than that of $AlEt_3$. Thus if $ZnEt_2$ is added to a $TiCl_3/AlEt_3$ system, it acts only as an efficient regulator[26] of molecular weight, without affecting catalyst activity.

There does exist some evidence for the role of $AlEt_3$ as a chain transfer agent. Keii and Soga[25] found that propylene polymerization continued for a considerable time even after the $AlEt_3$/*n*-heptane solu-

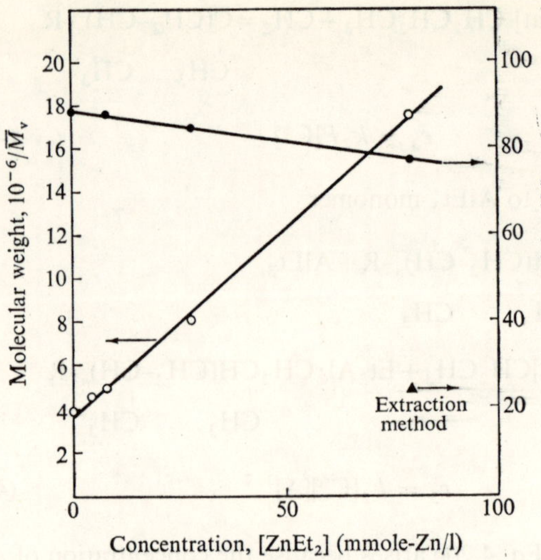

Fig. 4.18. Dependence of viscosity-average molecular weight and isotacticity on $ZnEt_2$ concentration.[26] (At 43.5°C in *n*-heptane)

tion was completely replaced by pure solvent (Fig. 4.19). The molecular weight of polymer produced after two hours was 100×10^5, which is $2 \sim 3$ times that formed with $TiCl_3$ and $AlEt_3$ ($[A]/[Ti] \simeq 1$), indicating that $AlEt_3$ probably acts as a chain transfer agent.

It is well known that hydrogen is the most important molecular weight regulator in practice. This will be discussed in the following section, together with the influence of hydrogen on reaction kinetics.

4.3.3 Effects of hydrogen on molecular weight and polymerization rate

Hydrogen is currently the most efficient industrial regulator of molecular weight. The mechanism shown below is supported by many authors

$$\text{cat-P} + H_2 \longrightarrow \text{cat-H} + \text{PH} \tag{4.21}$$

$$\text{cat-H} + \text{monomer} \longrightarrow \text{cat-P} \tag{4.21a}$$

where PH represents a terminated chain. Hoffman *et al.*[27] and Grieveson[28] consider that hydrogen acts as a simple transfer agent which should reduce the molecular weight without affecting the rate of polymerization, i.e. that Eq. 4.21*a* is a rapid process. This step is in

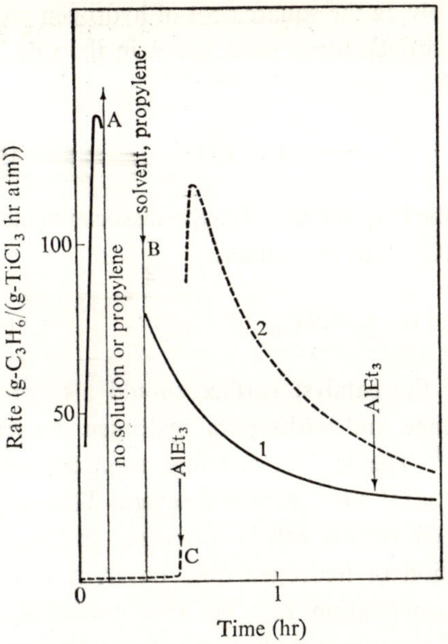

Fig. 4.19. Effect of $AlEt_3$ in the polymerization system. (AA-$TiCl_3$ = 2.5 g/l in 250 ml toluene. 1, Initially [$AlEt_3$] = 15 mmole/l. At A, the solution and propylene were removed, and pure solvent and propylene admitted (B). 2, Pure solvent was substituted for the $AlEt_3$ solution before introduction of propylene into the system. At C, $AlEt_3$ (15 mmole/l) was introduced in the presence of propylene.)

fact thought to be slow by Schindler,[29)] Natta *et al.*[30)] and Keii *et al.*,[31,32)] who confirmed a considerable decrease in polymerization rate in the presence of hydrogen for both ethylene and propylene, using $TiCl_3$ and $AlEt_3$ or $AlEt_2Cl$ catalysts.

Natta *et al.*[30)] obtained the following formula for the rate of propylene polymerization with $TiCl_3/AlEt_3$

$$R^0 - R^H = \frac{(P_H)^{1/2}}{k_2 + k_1(P_H)^{1/2}} \tag{4.22}$$

where R^H and R^0 are the rates in the presence and absence of hydrogen, respectively, and k_1, k_2 are constants. Putting $k_1/k_2 = \alpha$ and $k_1 = 1/R^0$ this can be converted to the more convenient form

$$R^H = \frac{R^0}{1 + \alpha(P_H)^{1/2}} \tag{4.22a}$$

The existence of a term involving the square root of hydrogen pressure suggests strongly that a hydrogen atom participates in the rate determining step, e.g.

$$\text{cat-P} + \text{H}_{ads} \longrightarrow \text{cat} + \text{PH} \tag{4.22b}$$

This assumes a pre-established equilibrium for the dissociative adsorption of hydrogen on the $TiCl_3$ catalyst surface.

$$\text{H}_2 \rightleftharpoons 2\text{H}_{ads} \tag{4.22c}$$

If this is indeed the case, the catalyst surface should also promote hydrogen/deuterium exchange and ortho/para hydrogen conversion, as pointed out by Hoeg.[33] The H_2/D_2 exchange reaction has been studied by Kojima and Keii,[31] who examined several $TiCl_3$/organometal catalyst systems. Their results will be discussed in Chapter 7.

If dissociative adsorption of hydrogen does occur, it will result in a lowering of the polymerization rate via two mechanisms; (i) reduction of the number of active centres available for polymerization, and (ii) temporary deactivation of polymerization centres following chain transfer.

Takahashi and Keii[32] examined the dependence of polymerization rate on hydrogen pressure using AA-grade $TiCl_3$ (specific surface area 47.7 m^2/g) and $AlEt_3$. In order to prevent the consumption† of hydrogen, the polymerization was carried out in a 1000 ml flask reactor, using 0.15 g $TiCl_3$ in 150 ml of *n*-heptane and with [A]/[Ti] = 1.0. No consumption of hydrogen and no change in propylene partial pressure could be detected by gas chromatography. A kinetic curve of the decay type was obtained both in the presence of hydrogen (0.1–30 cm Hg) and when hydrogen was not used. All the curves could be represented by

$$R = R_\infty + (\bar{R}_0 - R_\infty)e^{-k_3 t}$$

The observed values of k_3 at 43 °C are

† During polymerization (using the apparatus shown in Fig. 1.7) with 0.5 g-$TiCl_3$ in 250 ml of solvent (500 ml reactor), the propylene pressure increased from 30 to 62 cm Hg and the hydrogen pressure decreased from 39 to 6 cm Hg during the first 150 min.

$$P_H = 0 \text{ cm Hg} \quad k_3 = 1.48 + 3.2 \times 10^{-3}P$$
$$0.1 \quad k_3 = 1.47 + 3.7 \times 10^{-3}P \tag{4.23}$$
$$29 \quad k_3 = 1.49 + 3.0 \times 10^{-3}P$$

so it is apparent that the decay parameter k_3 is independent of hydrogen pressure. However, the stationary rate of polymerization is strongly dependent on hydrogen pressure as can be seen from Fig. 4.20. The relationship appears to be a Langmuir type

$$\frac{R^0_\infty - R^H_\infty}{R^0_\infty} = \frac{\alpha P_H^{1/2}}{1 + \alpha P_H^{1/2}} \tag{4.24a}$$

$$\frac{R^0_\infty - R^H_\infty}{R^0_\infty} = \frac{\alpha P_H}{1 + \alpha P_H} \tag{4.24b}$$

Eq. 4.24*a* corresponds to Eq. 4.22*a*, whereas the comparable form of Eq. 4.24*b* is

$$R^H_\infty = \frac{R^0_\infty}{1 + \alpha P_H} \tag{4.24c}$$

The results shown in Fig. 4.20 may be analysed in terms of the

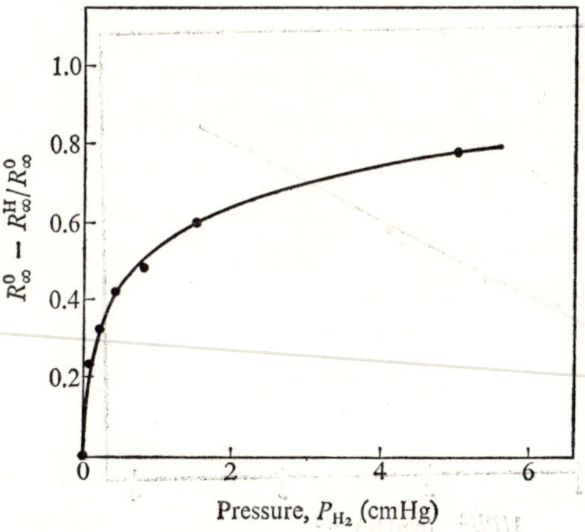

Fig. 4.20. Dependence of stationary rate on hydrogen pressure[32] (AA-$TiCl_3$/$EtAl_3$ at 43°C and 60 cmHg).

above equations by plotting $(R^0_\infty - R^H_\infty)/R^0_\infty$ against P_H and $P_H^{1/2}$, as in Fig. 4.21 and 4.22. It appears that the plot against $P_H^{1/2}$ gives a better fit. The observed values of α are shown in Table 4.8. The value of α increases with temperature according to the following equation.

$$\alpha = \alpha_0 e^{-13\text{kcal}/RT} \tag{4.24d}$$

These results demonstrate the role of adsorbed hydrogen atoms

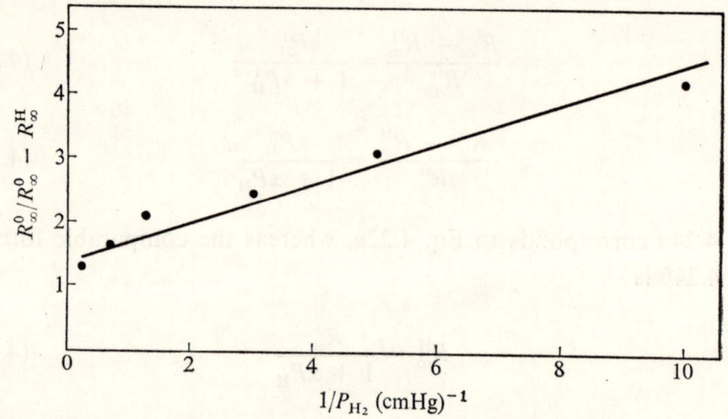

Fig. 4.21. Dependence of stationary rate on hydrogen pressure from Eq. 4.24 *b*.[32]

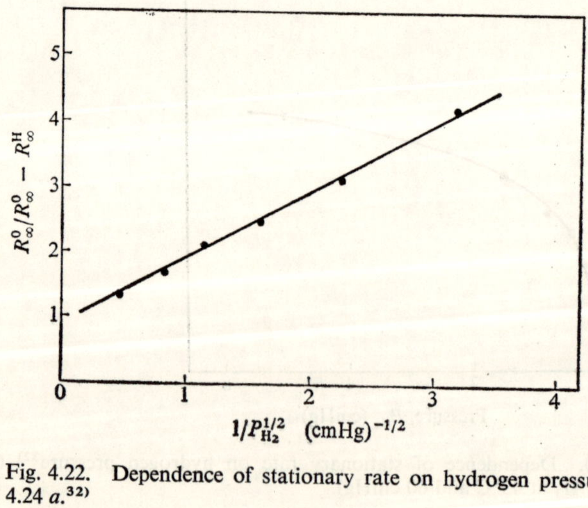

Fig. 4.22. Dependence of stationary rate on hydrogen pressure from Eq. 4.24 *a*.[32]

TABLE 4.8 Values of α

Temp. (°C)	43		32	E_α (kcal/mole) = 13
P(cmHg)	60	20	40	
α(cmHg)$^{-1/2}$	0.97	1.09	0.52	

on the catalyst surface. If the adsorption is Langmuir type, the surface coverage of hydrogen atoms can be represented by Eq. 3.14, and the surface concentration will then be

$$[\mathrm{H}] = [\mathrm{S}]\frac{(K_{\mathrm{H}}P_{\mathrm{H}})^{1/2}}{1 + (K_{\mathrm{H}}P_{\mathrm{H}})^{1/2}} \tag{4.25}$$

where [S] is the concentration of surface sites. The effective concentration of active centres, $[\mathrm{C}]_{\mathrm{eff}}$, available for polymerization centre formation will then be

$$[\mathrm{C}]_{\mathrm{eff}} = [\mathrm{S}] - [\mathrm{H}] = \frac{[\mathrm{S}]}{1 + (K_{\mathrm{H}}P_{\mathrm{H}})^{1/2}}$$

which gives a resultant rate of polymerization in the presence of hydrogen of

$$R^{\mathrm{H}} = \frac{R^0}{1 + (K_{\mathrm{H}}P_{\mathrm{H}})^{1/2}} \tag{4.25a}$$

thus accounting for the form of the experimental equation.

However, the adsorption constant may be expressed as $K_{\mathrm{H}} = K_{\mathrm{H}}^0 \exp(E_{\mathrm{H}}/2RT)$ where the energy (heat) of adsorption, $E_{\mathrm{H}} > 0$, represents the difference between the energy of two adsorbed hydrogen atoms and the groundstate of a gaseous hydrogen molecule. Experimentally, the energy of α (Eq. 4.22*a*) is known to be negative, and thus the $K_{\mathrm{H}}^{1/2}$ term in Eq. 4.25*a* cannot be regarded as equivalent to α, refuting the simple theory of retardation of polymerization by adsorbed hydrogen atoms. Takahashi and Keii considered the second possibility previously referred to, using the following model

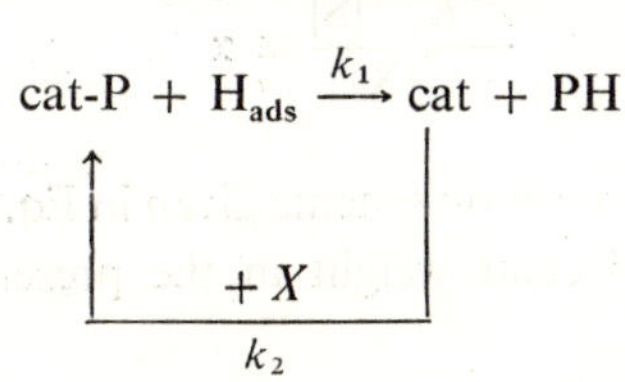

(4.26)

For stationary state polymerization, the rates of transfer (deactivation) and recovery of polymerization centres will be equal

$$k_1[\mathrm{H}][\mathrm{C}^*] = k_2[\mathrm{C}][\mathrm{X}] \tag{4.26a}$$

If it is assumed that the surface concentration of adsorbed hydrogen atoms is negligible compared with the total number of active centres, i.e.

$$[\mathrm{H}] \ll [\mathrm{C}^*] + [\mathrm{C}] \equiv [\mathrm{S}] \tag{4.26b}$$

then the Langmuir adsorption equation may be approximated as follows

$$[\mathrm{H}] \simeq [\mathrm{S}]\,(K_{\mathrm{H}}P_{\mathrm{H}})^{1/2} \tag{4.26c}$$

From the above equation and Eq. 4.26*a*, we have

$$[\mathrm{C}^*] = \frac{[\mathrm{S}]}{1 + \dfrac{k_1}{k_2[\mathrm{X}]}[\mathrm{H}]}$$

$$= \frac{[\mathrm{S}]}{1 + \left(\dfrac{k_1[\mathrm{S}]K_{\mathrm{H}}^{1/2}}{k_2[\mathrm{X}]}\right)P_{\mathrm{H}}^{1/2}} \tag{4.26d}$$

The stationary rate of polymerization in the presence of hydrogen can then be expressed as

$$R_{\infty}^{\mathrm{H}} = \frac{R_{\infty}^{0}}{1 + \left(\dfrac{k_1[\mathrm{S}]K_{\mathrm{H}}^{1/2}}{k_2[\mathrm{X}]}\right)P_{\mathrm{H}}^{1/2}} \tag{4.26e}$$

which is compatible with the experimental results (Eq. 4.22*a*) provided that

$$\frac{k_1K_{\mathrm{H}}^{1/2}[\mathrm{S}]}{k_2[\mathrm{X}]} \equiv \alpha \tag{4.26f}$$

On the basis of the reaction scheme given in Eq. 4.26, the stationary value for polymer molecular weight in the presence of hydrogen is given by

$$M_{\infty}^{\text{H}} = \frac{R_{\infty}^{\text{H}}}{R_t^0 + k_1[\text{C}^*][\text{H}]} \tag{4.26g}$$

where the stationary polymerization rate $R_{\infty}^{\text{H}} = k_p[\text{M}][\text{C}^*]$, the total transfer rate independent of hydrogen $R_t^0 = k_t[\text{C}^*]$, and $k_1[\text{C}^*][\text{H}]$ represents the transfer due to adsorbed hydrogen atoms. This equation may be rewritten† as

$$M_{\infty}^{\text{H}} = \frac{M_{\infty}^0}{1 + \dfrac{k_1 K_{\text{H}}^{1/2}[\text{S}]}{k_t} P_{\text{H}}^{1/2}} \tag{4.26h}$$

with

$$M_{\infty}^0 = \frac{k_p[\text{M}]}{k_t} = \frac{R_{\infty}^0}{k_t[\text{S}]}$$

This final equation is compatible with the experimental results of Takahashi and Keii[32] and also with those of Kojima and Keii,[31] who proposed the experimental formula Eq. 4.27.

$$M_{\infty}^{\text{H}} = \frac{M_{\infty}^0}{1 + \beta P_{\text{H}}^{1/2}} \tag{4.27}$$

The results of Kojima and Keii are shown in Fig. 4.23 and 4.24, which support Eq. 4.27 and also indicate that

$$R_t^0 = R_t^{0,0} + k_{\text{tr}}^{\text{P}} P \tag{4.28}$$

which suggests that the transfer reaction is partly attributable to propylene monomer. From Eq. 4.27 and 4.28 we have

$$\begin{aligned} \frac{R_{\infty}^0}{M_{\infty}^{\text{H}}} &= \frac{R_{\infty}^0}{M_{\infty}^0}(1 + \beta P_{\text{H}}^{1/2}) \\ &= R_t^0(1 + \beta\, P_{\text{H}}^{1/2}) \\ &= \underbrace{R_t^{0,0}(1 + \beta\, P_{\text{H}}^{1/2})}_{\text{intercept}} + \underbrace{k_{\text{tr}}^{\text{P}}(1 + \beta\, P_{\text{H}}^{1/2})P}_{\text{slope}} \end{aligned} \tag{4.28a}$$

† The [S] in the denominator can be eliminated, since $k_1[\text{C}^*][\text{H}]$ should strictly be replaced by the more correct form $k_1[\text{C}^*]\,\theta_{\text{H}}$ (see 3.1.2).

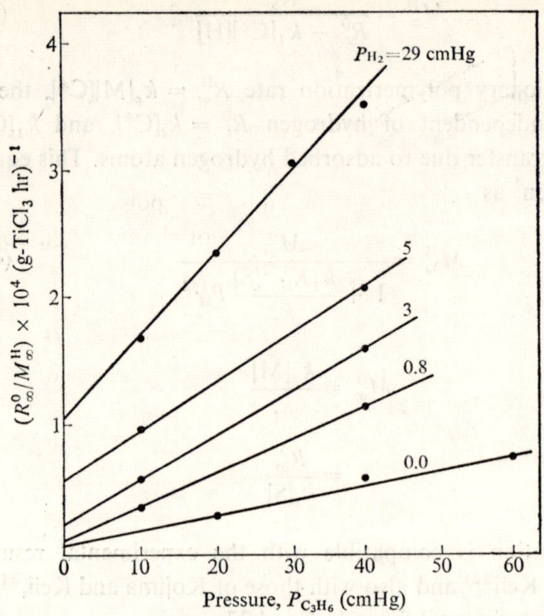

Fig. 4.23. Relationship between R^0_∞/M^H_∞ and propylene pressure at various hydrogen pressures.[31] (AA-TiCl$_3$ = [AlEt$_3$] = 1 mmole/150 ml of toluene at 43°C.)

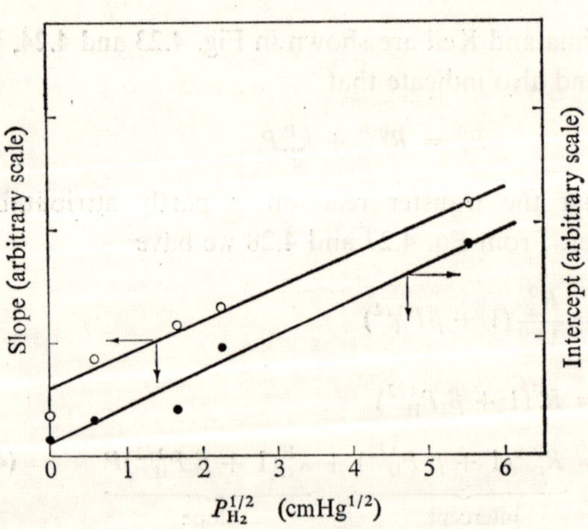

Fig. 4.24. Intercepts and slopes from Fig. 4.23 *vs.* hydrogen pressure to the power 1/2.[31]

indicating similar dependences of the slope and intercept on hydrogen pressure, as illustrated in Fig. 4.24.

The experimental results can thus be explained in terms of the reaction scheme in Eq. 4.26, though this does involve a conventional step with the species X involved in polymerization centre recovery. There is some experimental evidence to indicate that X corresponds to $AlEt_3$ and a propylene monomer. Thus, the polymerization rate, depressed by the addition of hydrogen, did not change when both the hydrogen and the $AlEt_3$ were rapidly removed from the system but recovered on readmittance of $AlEt_3$, as shown in Fig. 4.25. When the hydrogen alone was removed, recovery was rather slow (Fig. 4.26). These results are compatible with the initial kinetic behavior of polymerization systems, explained in terms of the formation of polymerization centres by both $AlEt_3$ and propylene monomer.

The changes in polymerization rate and molecular weight which occur during propylene polymerization with $TiCl_3/AlEt_2Cl$ are clearly

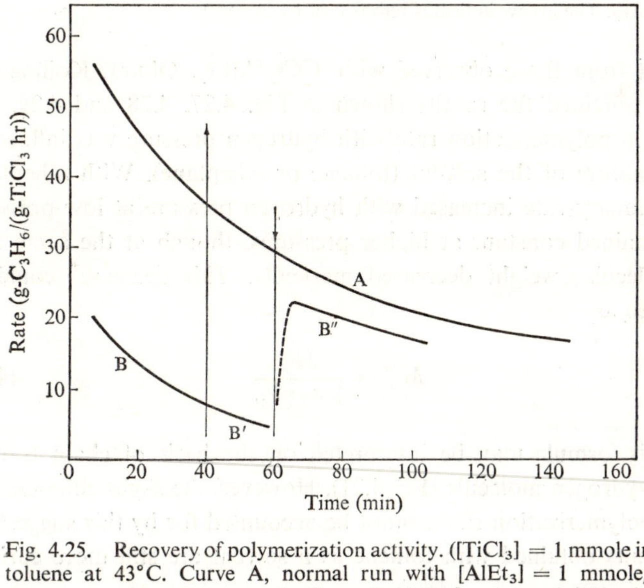

Fig. 4.25. Recovery of polymerization activity. ($[TiCl_3]$ = 1 mmole in 150 ml toluene at 43°C. Curve A, normal run with $[AlEt_3]$ = 1 mmole/150 ml toluene, $P = 30$ cmHg, $P_H = 0$; Curve B, $[AlEt_3]$ = 1 mmole/150 ml toluene, $P = 30$ cmHg, $P_H = 29$ cmHg; curve B′, $[AlEt_3] = 0$, $P = P_H = 0$; curve B″, $[AlEt_3]$ = 1 mmole/150 ml toluene, $P = 30$ cmHg, $P_H' = 0$. The upward arrow indicates removal of H_2, and the downward one the readmittance of $AlEt_3$).

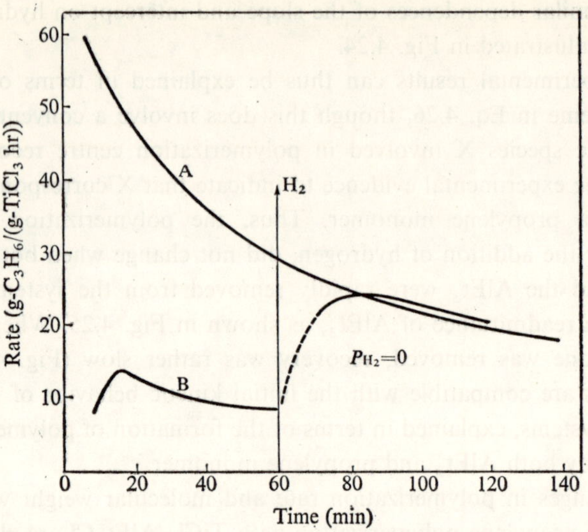

Fig. 4.26. Recovery of polymerization activity on removal of hydrogen. (Curve A, normal run with $[AlEt_3] = 1$ mmole/150 ml toluene, $P = 30$ cmHg, $P_H = 0$ at 43°C; curve B, as above in the presence of hydrogen, $P_H = 29$ cmHg. The arrow indicates removal of hydrogen.)

different from those observed with $TiCl_3/AlEt_3$. Okura, Kojima and Keii[34)] obtained the results shown in Fig. 4.27, 4.28 and 4.29. The change in polymerization rate with hydrogen pressure was influenced by the nature of the solvent (toluene or *n*-heptane). With *n*-heptane, the stationary rate increased with hydrogen pressure at low pressures and remained constant at higher pressures, though at the same time the molecular weight decreased markedly. This decrease† could be expressed as

$$M_\infty^{\mathrm{H}} = \frac{M_\infty^0}{1 + \gamma P_{\mathrm{H}}} \tag{4.29}$$

This formula may be interpreted on the basis of chain transfer by the hydrogen molecule (Eq. 4.21). However, the slight enhancement of the polymerization rate cannot be accounted for by this suggestion. The results obtained with toluene as a solvent are still more curious.

† The apparent linear relationship between $(M_\infty^{\mathrm{H}})^{-1}$ and $P_{\mathrm{H}}^{1/2}$ which can be obtained from the data in Fig. 4.29 is clearly not a valid one, since the intercept has a negative value.

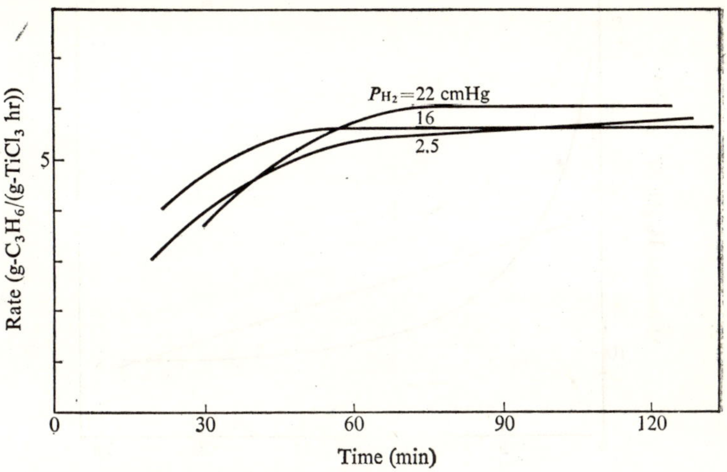

Fig. 4.27. Kinetic curve for $TiCl_3/AlEt_2Cl/H_2$ system at 45°C.[34] ([$TiCl_3$] = 13 mmole/l, [$AlEt_2Cl$] = 16 mmole/l in 250 ml *n*-heptane with P = 40 cmHg).

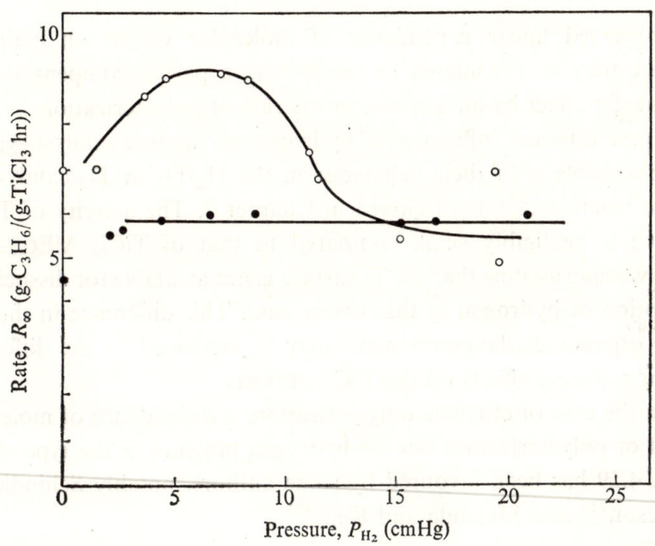

Fig. 4.28. Solvent effects on the hydrogen pressure dependence of the stationary rate for the $TiCl_3/AlEt_2Cl/H_2$ system.[34] (Conditions as in Fig. 4.27. *n*-Heptane ●, and toluene ○.)

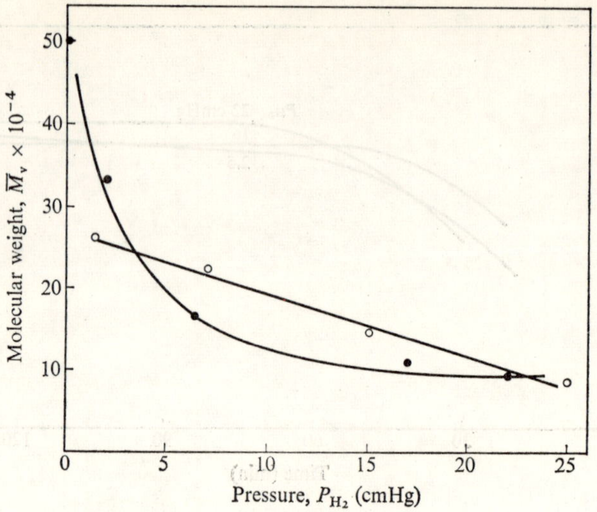

Fig. 4.29. Hydrogen pressure dependence of viscosity-average molecular weight in different solvents.[34] (Conditions as in Fig. 4.28.)

The observed linear dependence of molecular weight on hydrogen pressure may be considered to result from a partial compensation of the transfer effect by an increase in the rate of polymerization.

These different influences of hydrogen on the two catalyst systems are compatible with their behaviour in the H_2/D_2 or D_2/catalyst exchange reactions,[35] as discussed in Chapter 7. The activity of $TiCl_3/AlEt_2Cl$ is negligibly small compared to that of $TiCl_3/AlEt_3$, from which we may assume that the Ti surface is not available for dissociative adsorption of hydrogen in the former case. This difference in the role of the organometallic components may be explained by the difference in their reducing effects on the $TiCl_3$ surface.

In the case of ethylene polymerization, a dependence of molecular weight or polymerization rate on hydrogen pressure of the type shown in Eq. 4.29 has been favoured by some authors, notably Schindler[29], Grieveson[28] and Matsuda and Keii.[35]

References

1. G. Natta and I. Pasquon, *Advan. Catalysis*, **11**, 1 (1959).
2. J. P. Luongo, *Anal. Chem.*, **33**, 1816 (1961).
3. A.P. Firsov, B.N. Kashpopov and N.M. Chirkov, *Polymer Sci. USSR*, **6**, 403 (1965).
4. K. Soga and T. Keii, *J. Polymer Sci.*, part A1, **4**, 2429 (1966).
5. G. Natta, P. Pino, M. Mazzanti and P. Longi, *Gazz. Chim. Ital.*, **87**, 570 (1959); G. Natta and I. Pasquon, *Advan. Catalysis*, **11**, 1 (1959).
6. F. Dawans and P. Teyssie, *Bull. Soc. Chim. France*, **1963**, 2376.
7. T. Keii, K. Soga, K. Go and M. Kojima, *J. Polymer Sci.*, **C23**, 453 (1968).
8. J. Boor Jr., *J. Polymer Sci.*, **C1**, 237 (1963).
9. D.O. Jordan, *The Stereochemistry of Macromolecules* (ed. A.D. Ketley), vol. 1, p. 31, Dekker, 1967.
10. N.G. Gaylord and H. Mark, *Linear and Stereoregular Addition Polymers*, p. 211, Interscience, 1959.
11. F. Danusso, *J. Polymer Sci.*, **C4**, 149 (1964).
12. G. Natta, I. Pasquon, A. Zambelli and G. Gatti, *J. Polymer Sci.*, **51**, 387 (1961).
13. T. Keii, Y. Doi and I. Okura, unpublished paper.
14. A.K. Ingberman, I.J. Levine and R.J. Turbett, *J. Polymer Sci.*, part A1, **4**, 2781 (1966).
15. M.N. Berger and B.M. Grieveson, *Makromol. Chem.*, **83**, 80 (1964).
16. D.F. Hoeg and S. Liebman, *Ind. Eng. Chem. Process Design Develop.*, **1**, 120 (1962).
17. G. Natta, G. Mazzanti, D. De Luca, U. Giannini and F. Bandini, *Makromol. Chem.*, **83**, 54 (1964).
18. T. Keii and K. Soga, *Kogyo Kagaku Zasshi* (Japanese), **70**, 1876 (1967); D.H. Hoeg, ref. 9, p. 120.
19. F.S. Dyachkovskii, A.K. Shilova and E. Shilov, *J. Polymer Sci.*, **C16**, 2333 (1967).
20. D.F. Hoeg and S. Liebman, *Belg. Pat.* 589260 (1960); ref. 9, p. 129.
21. Y. Murayama and T. Keii, *Shokubai* (Japanese), **5**, 247 (1963).
22. J.B. Kinsinger, presented at Meeting of Amer. Chem. Soc., (1957); J. Phys. Chem., **63**, 2002 (1959).
23. D.F. Hoeg, ref. 9, p. 120.
24. G. Natta, E. Giachetti, I. Pasquon and P. Pino, *Chem. Ind.* (*Milan*), **42**, 1091 (1960).
25. T. Keii, K. Soga and N. Saiki, *J. Polymer Sci.*, **C16**, 1507 (1967).
26. D.F. Hoeg, ref. 9, p. 120.
27. A.S. Hoffman, B.A. Fries and P.C. Condit, *J. Polymer Sci.*, **C4**, 109 (1963).
28. B.M. Grieveson, *Makromol. Chem.*, **84**, 93 (1965).
29. A. Schindler, *J. Polymer Sci.*, **C4**, 81 (1963).
30. G. Natta, G. Mazzanti, P. Longi and F. Bernadini, *Chim. Ind.* (*Milan*), **41**, 519 (1959).
31. M. Kojima and T. Keii, unpublished paper; M. Kojima, *M. Eng. Thesis, Tokyo Inst. Technology*, 1968.
32. A. Takahashi and T. Keii, presented at Japan Chem. Soc. Annual Meeting, 1966; A. Takahashi, *M. Eng. Thesis*, *Tokyo Inst. Technology*, 1966.
33. D.F. Hoeg, ref. 9, p. 123.
34. I. Okura, A. Kojima, K. Soga and T. Keii, *J. Polymer Sci.*, part A1, **8**, 2717 (1970).
35. T. Matsuda and T. Keii, unpublished paper; T. Matsuda, *Dr. Eng. Thesis*, *Tokyo Inst. Technology*, 1967.

Chapter 5

Ethylene polymerization

5.1 KINETICS OF ETHYLENE POLYMERIZATION WITH $TiCl_3/AlEt_3$ AND $TiCl_3/AlEt_2Cl$

Ethylene polymerization using Ziegler catalysts based on $TiCl_4$ has been widely studied, but work with $TiCl_3$ based catalysts has been considerably less comprehensive. In general, the polymerization of ethylene with $TiCl_3$ and $AlEt_3$ or $AlEt_2Cl$ is kinetically similar to propylene polymerization with the same catalysts, though there are differences of detail.

The rate of ethylene polymerization changes with time even using $TiCl_3/AlEt_2Cl$ catalyst systems, as pointed out by Berger and Grieveson,[1)] and as in the case of propylene, it is therefore preferable to use differential rates of polymerization.

Matsuda and Keii[2)] examined constant pressure ethylene polymerization with $TiCl_3$ (AA-grade, surface area 40 m^2/g) and $AlEt_3$, and found that the rate was almost constant during the initial stages ($t < 15$ min) after which it decreased gradually with time until the stationary value was attained. Fig 5.1 shows ethylene consumption *vs.* time at 40 °C and 40 cm Hg ethylene partial pressure in the presence and absence of hydrogen and nitrogen. The figure shows that nitrogen has no apparent effect on the rate, but that hydrogen depresses the rate considerably. Schindler[3)] has previously reported some depression of the rate in the presence of nitrogen, using a $TiCl_4/AlEt_2Cl$ catalyst system. Matsuda and Keii[2)] examined the applicability of a mixed first and second order dependence of initial rate, R_0, on ethylene pressure proposed by Schindler and obtained the results shown in Fig. 5.2.

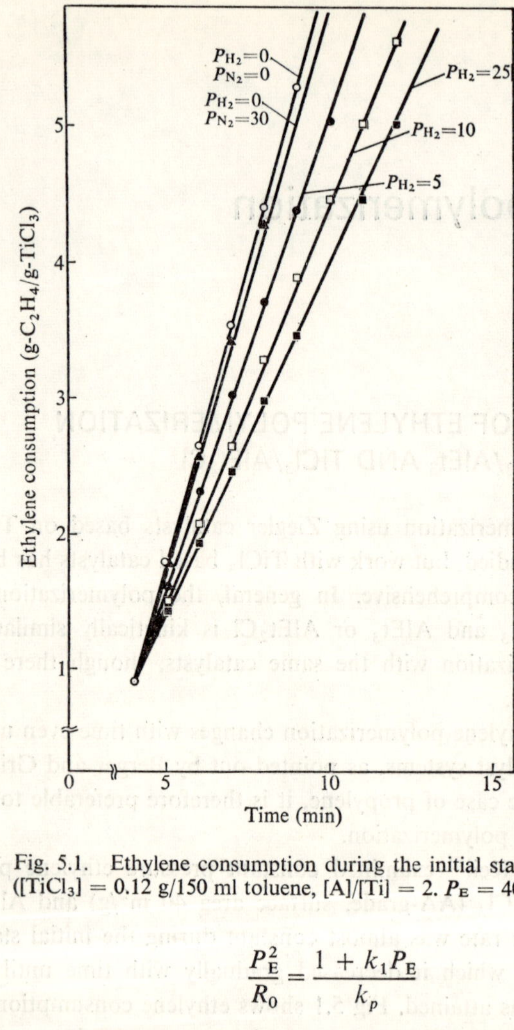

Fig. 5.1. Ethylene consumption during the initial stages of polymerization ($[TiCl_3]$ = 0.12 g/150 ml toluene, [A]/[Ti] = 2. P_E = 40 cmHg at 40°C.)

$$\frac{P_E^2}{R_0} = \frac{1 + k_1 P_E}{k_p} \tag{5.1}$$

The plot of the above equation in Fig. 5.2 appears linear, but the validity of Eq. 5.1 has been questioned on the basis of the negative value of the intercept.

The gradual decrease of the polymerization rate with time has been examined at various pressures and temperatures and the observed kinetic curves are shown in Fig. 5.3 and 5.4. The rate dependence could be expressed by Eq. 5.2 (Eq. 2.7), which is also applicable to propylene polymerization.

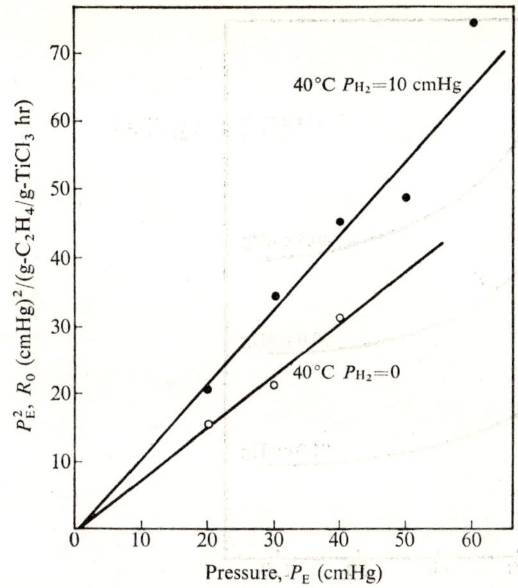

Fig. 5.2. Pressure dependence of the initial rate of polymerization.

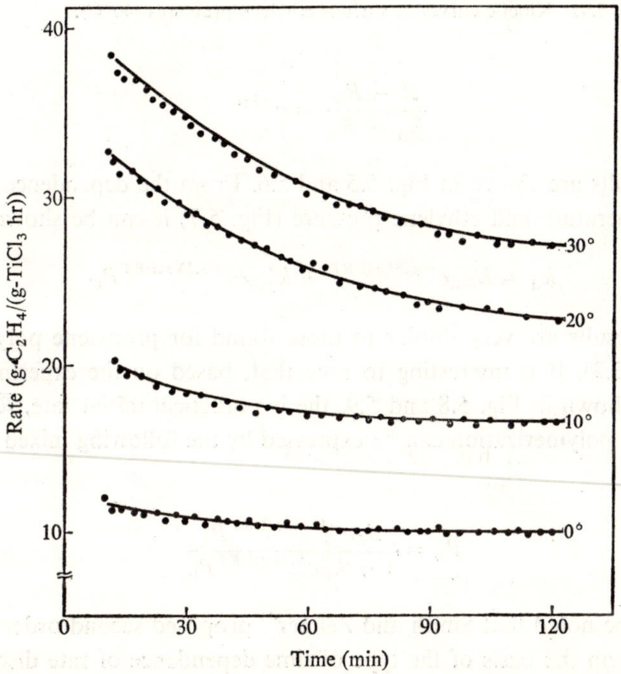

Fig. 5.3. Kinetic curves at various temperatures. (P_E = 40.5 cmHg, [A]/[Ti] = 2.)

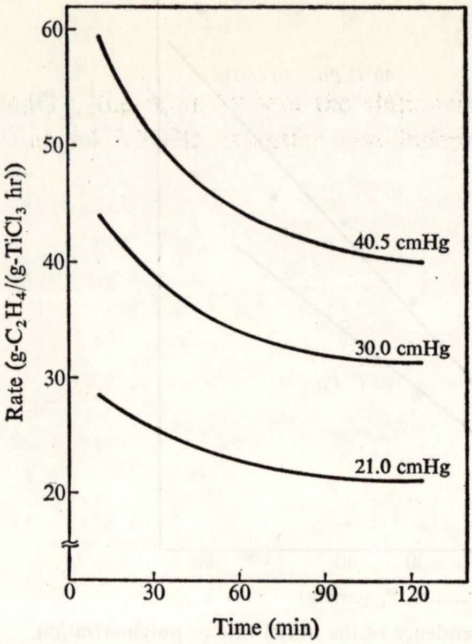

Fig. 5.4. Kinetic curves at various ethylene pressures (40°C).

$$\frac{R - R_\infty}{\bar{R}_0 - R_\infty} = e^{-k_3 t} \tag{5.2}$$

The results are shown in Fig. 5.5 and 5.6. From the dependence of k_3 on temperature and ethylene pressure (Fig. 5.7) it can be shown that

$$k_3 = k'_{3,0}e^{-4.8\text{kcal}/RT} + k''_{3,0}e^{-10.5\text{kcal}/RT}P_\text{E} \tag{5.3}$$

These results are very similar to those found for propylene polymerization (2.3). It is interesting to note that, based on the experimental results shown in Fig. 5.8 and 5.9, the hypothetical initial rate, $\bar{R}_0$, for ethylene polymerization can be expressed by the following mixed order rate law

$$\bar{R}_0 = \frac{A_0 e^{-11.8\text{kcal}/RT} P_\text{E}^2}{1 + k_\text{E}^0 e^{3.3\text{kcal}/RT} P_\text{E}} \tag{5.4}$$

It may be noted that Smith and Zelmer[4)] proposed second order deactivation on the basis of the type of time dependence of rate discussed

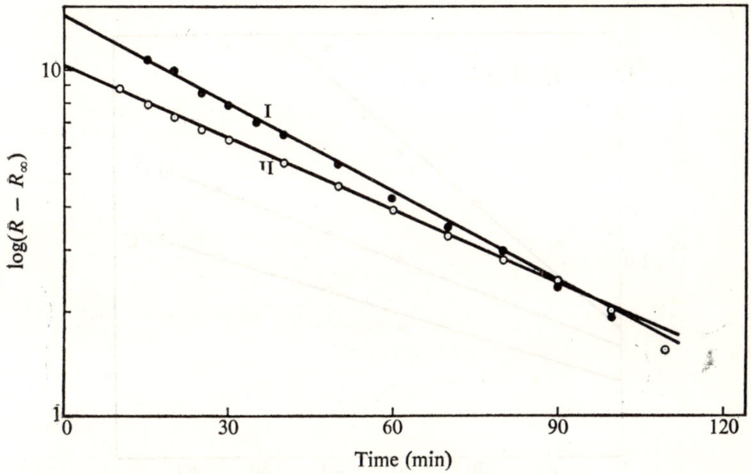

Fig. 5.5. Time dependence of polymerization rate from Eq. 5.2. (I, P_E = 40.5 cmHg, [A]/[Ti] = 2, 20°C; II, 30°C.)

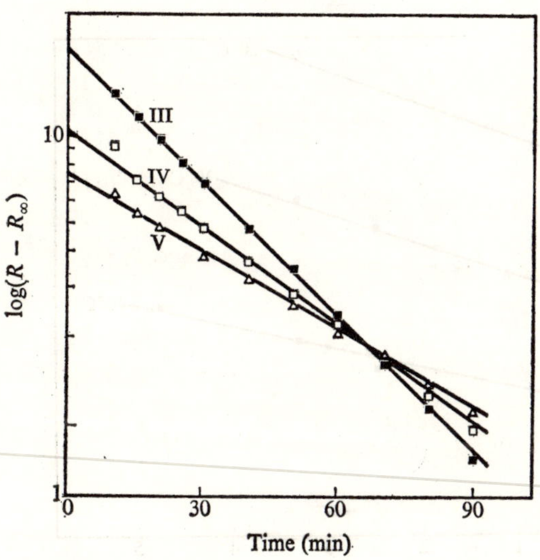

Fig. 5.6. Time dependence of polymerization rate from Eq. 5.2. (III, P_E = 40.5 cmHg, [A]/[Ti] = 2, 40°C; IV, P_E = 30 cmHg; V, P_E = 20 cmHg.)

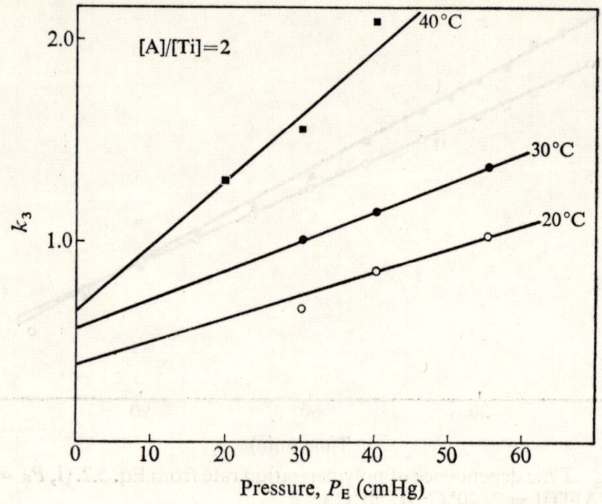

Fig. 5.7. Pressure dependence of k_3 at various temperatures.

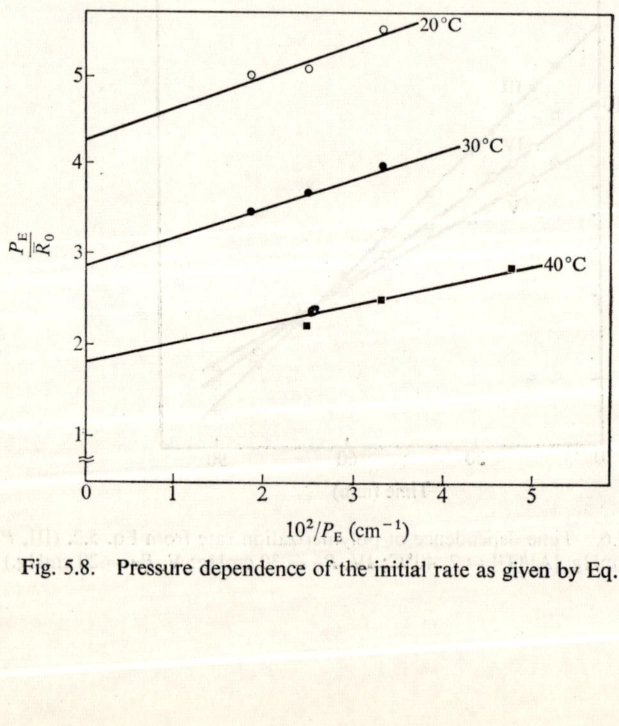

Fig. 5.8. Pressure dependence of the initial rate as given by Eq. 5.4.

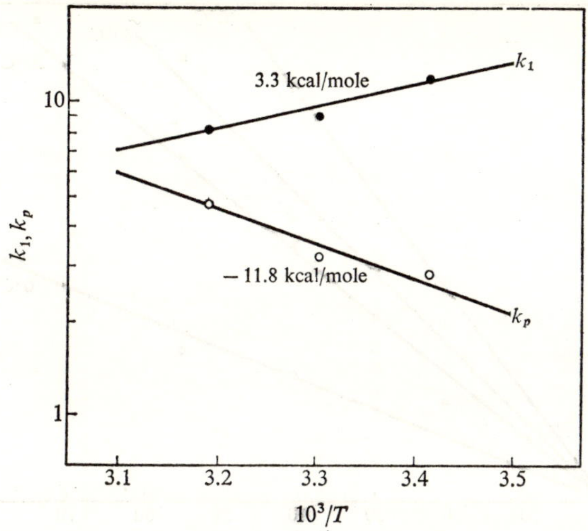

Fig. 5.9. Temperature dependence of the constant of Eq. 5.4.

in 3.3. This has also been favoured by Schindler[3,5] for curves obtained with $TiCl_4$ and $AlEt_2Cl$ or $ZnEt_2$.[6] These will be discussed in 5.3.

The observed stationary rates of ethylene polymerization can be expressed by a first order rate law with respect to pressure, as in Fig. 5.10, from which the stationary state activation energy can be obtained as 6.1 kcal/mole. The dependence of the stationary rate and the hypothetical initial rate on $AlEt_3$ concentration are the same as those found for propylene, as can be seen from Fig. 5.11 and 5.12, and the stationary rate can then be expressed as

$$R_\infty = Ae^{-6.1\text{kcal}/RT} P_E \left(\frac{K[A]}{1 + K[A]} \right) \tag{5.5}$$

The independence of the decay parameter, k_3, on $AlEt_3$ concentration has been confirmed over a wide concentration range (Fig. 5.13), and this is important in considering the mechanism of the rate decrease during polymerization (3.3).

The catalyst system $TiCl_3/AlEt_2Cl$ gives acceleration type kinetic curves with ethylene, as with propylene. The effect of stepwise temperature changes on the rate is shown in Fig. 5.14, and the dependence of the stationary rate on $AlEt_2Cl$ concentration at various temperatures

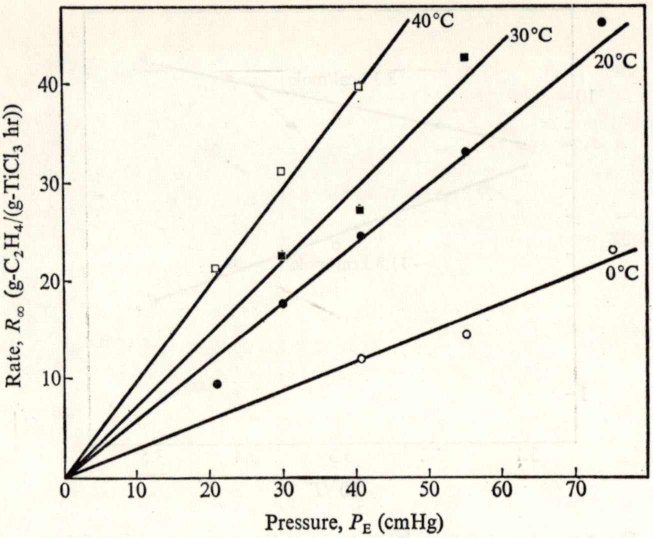

Fig. 5.10. Pressure dependence of the stationary rate at various temperatures.

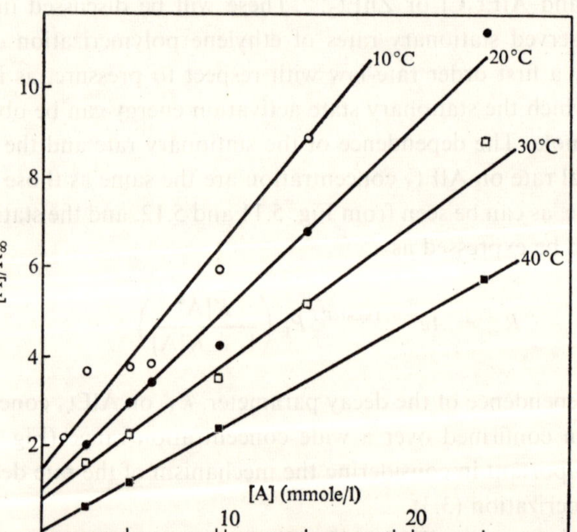

Fig. 5.11. Concentration dependence of the stationary rate at various temperatures as given by Eq. 5.5.

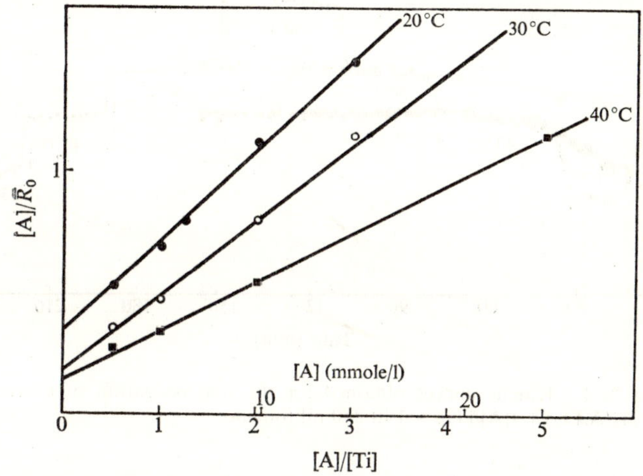

Fig. 5.12. Concentration dependence of the hypothetical inital rate at various temperatures.

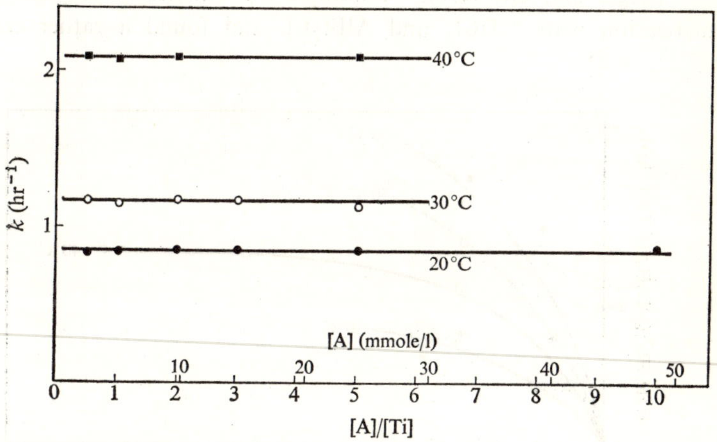

Fig. 5.13. Relationship between k_3 and $AlEt_3$ concentration at various temperatures (P_E = 40.5 cmHg)

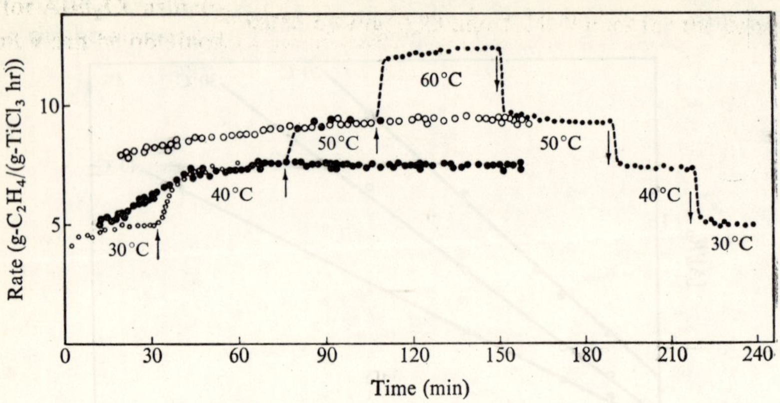

Fig. 5.14. Kinetic curves obtained for the polymerization of ethylene by $TiCl_3/AlEt_2Cl$. ([A]/[Ti] = 1 in 250 ml toluene.)

is shown in Fig. 5.15. The dependence is a Langmuir type, as can be seen from Fig. 5.16, which is confirmed by the results of Schnecko *et al.*[7-9] The adsorption constant in this case could be expressed as

$$K = K_0 e^{11 \sim 13 \mathrm{kcal}/RT} \tag{5.6}$$

Berger and Grieveson[1] examined the kinetics of ethylene polymerization with γ-$TiCl_3$ and $AlEt_2Cl$, and found a rather curious

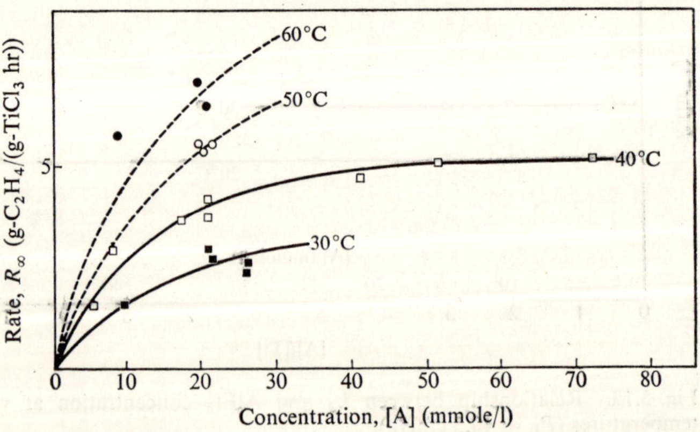

Fig. 5.15. Dependence of the stationary rate on $AlEt_2Cl$ concentration at various temperatures.

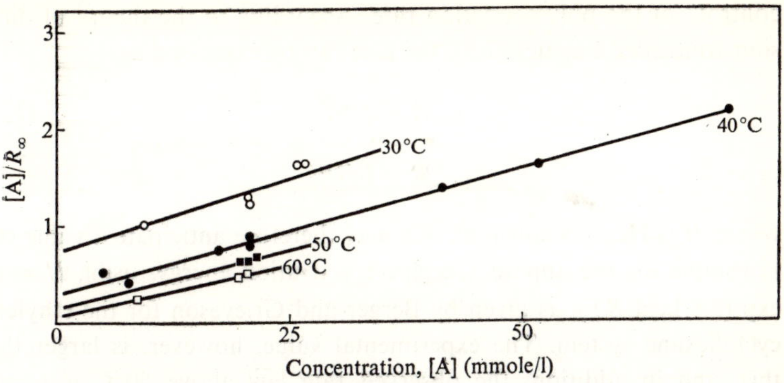

Fig. 5.16. Dependence of stationary rate on $AlEt_2Cl$ concentration at various temperatures as given by Eq. 5.5. (Data from Fig. 5.15.)

temperature dependence of the stationary rate (Fig. 5.17). The activation energy below 50 °C is 13.5 kcal/mole, which may be compared with the value of 7 kcal/mole obtained by Matsuda and Keii,[2)] while above 50 °C the apparent value of the activation energy is negative. It was supposed by Berger and Grieveson that this was due to "diffusion

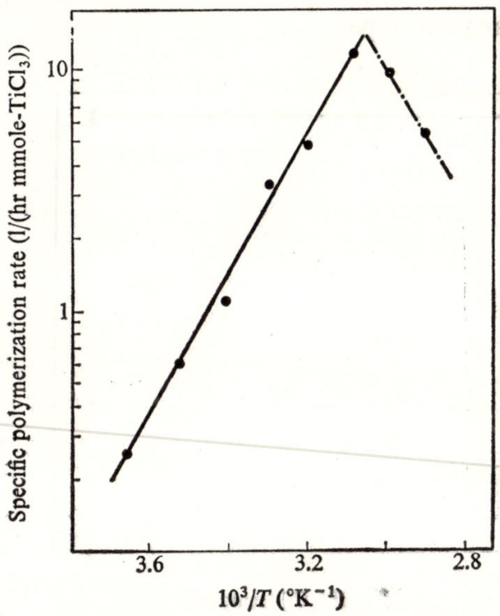

Fig. 5.17. Arrhenius plot of ethylene polymerization rates.[1)] ($[TiCl_3]$ = 5.25 mmole/l, $[AlEt_2Cl]/[Ti]$ = 1, P = 760 mmHg.)

control" of the polymerization rate. According to the theory of diffusion controlled kinetics (1.3), the rate can be expressed as

$$R = \beta[\mathrm{M}]_0 \qquad (1.8b)$$

or

$$R = \beta H P_{\mathrm{E}}$$

where H is Henry's constant. We may therefore anticipate a value of 4 kcal/mole for the apparent negative activation energy, using $H = H_0 \exp(3.8\ \mathrm{kcal}/RT)$, as given by Berger and Grieveson for the ethylene/cyclohexane system. The experimental value, however, is larger than this, and in addition, the observed rate law above 50°C is second order with respect to ethylene pressure, which is not compatible with the diffusion control hypothesis (Eq. 5.6). The dependence of polymerization rate on ethylene pressure over the range 10–70 cm Hg has been summarized by Berger and Grieveson as follows

$$R_\infty = \alpha P_{\mathrm{E}}^n \qquad (5.7)$$

where $n = 1.1$ (30°C), 1.0 (40°C) and 2.2 (60°C). The dependence of the rate on $AlEt_2Cl$ concentration is shown in Fig. 5.18, and appears to be a higher order Langmuir type

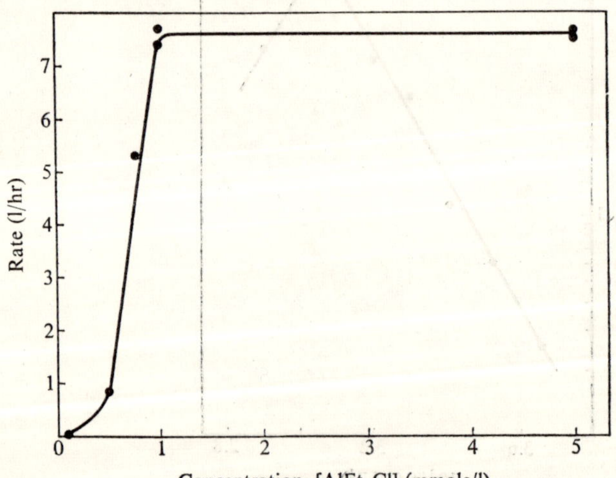

Fig. 5.18. Effect of $AlEt_2Cl$ concentration on polymerization rate.[1] ($[TiCl_3]$ = 5 mmole/l, P = 760 mm Hg at 40°C.)

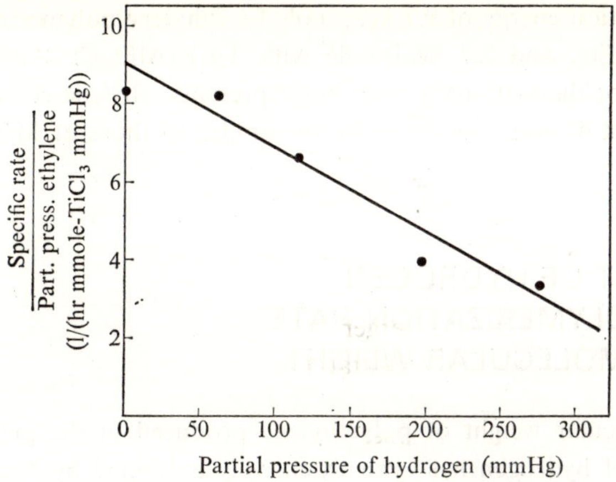

Fig. 5.19. Effect of hydrogen of ethylene polymerization rate. ($[\gamma\text{-TiCl}_3] = [\text{AlEt}_2\text{Cl}] = 5.25$ mmole/l, $P = 760$ mm Hg at 40°C.)

$$R = R^0\left(\frac{K[\text{A}]^s}{1 + K[\text{A}]^s}\right) \tag{5.8}$$

with $s \geq 1$, at least for these results. The effect of hydrogen on the rate was also examined (Fig. 5.19).

Lanovskaya *et al.*[10,11] investigated ethylene polymerization with α- or β-$TiCl_3$ and $AlEt_3$, $AlEt_2Cl$, $ZnEt_2$, $MgEt_2$, LiEt and Al(i-Bu)$_3$. The activation energies found for $AlEt_3$ and $AlEt_2Cl$ are shown in Table 5.1. Fukui, Kagiya *et al.*[12] investigated the $TiCl_3/AlEt_3$ catalyst system at high pressure (42 kg/cm^2) and constant volume, and obtained a value of 1.6 kcal/mole for the activation energy over the temperature range 0–100°C. The rate law obtained by Schnecko *et al.*[9]

$$R = kP_{\text{E}}\left(\frac{K[\text{A}]}{1 + K[\text{A}]}\right) \tag{5.9}$$

TABLE 5.1 Activation energy of ethylene polymerization[10,11]†

Catalyst \ [A]/[Ti]	1.0	2.0	5.5	10.0	10.7
$AlEt_3$		7.3			8.4
$AlEt_2Cl$	7.5	8.4	8.2	8.0	

† α-, β-$TiCl_3$, ethylene pressure 200–400 mm Hg.

gave an activation energy of 9.7 kcal/mole for ethylene polymerization with $TiCl_3/AlEt_3$, and 8.7 kcal/mole with $TiCl_3/AlEt_2Cl$. Although their rate is not the stationary rate, but represents ethylene consumption in the first 45 min, these results are similar to those of Matsuda and Keii.[2)]

5.2 EFFECT OF HYDROGEN ON POLYMERIZATION RATE AND MOLECULAR WEIGHT

The molecular weight of polyethylene produced in the presence and absence of hydrogen has been extensively examined by Matsuda and Keii,[2)] using AA-grade $TiCl_3$ with $AlEt_3$ or $AlEt_2Cl$. Eq. 5.10 was used to calculate the molecular weight from the intrinsic viscosity, η, in decalin at 135 °C.

$$[\eta] = 5.46 \times 10^{-4} \overline{M}_v^{0.725} \quad (5.10)$$

Fig. 5.20 shows the observed linear relationship between the reciprocals of molecular weight and polymerization time. After two hours the molecular weight can be regarded as nearly stationary. The observed relationship between the stationary molecular weight and the $AlEt_3$ concentration (Fig. 5.21) can be expressed as

$$\frac{1}{\overline{M}_v} \propto [A]^{1/2} \quad (5.11)$$

The dependence of molecular weight on ethylene pressure is shown in Fig. 5.22, which may be summarized as

$$\frac{1}{\overline{M}_v} \propto P_E \quad (5.12)$$

These results are very similar to those found during propylene polymerization.

Matsuda and Keii calculated the mean lifetime, $\bar{L}$, of the growing polymer chains by the method of Natta *et al.*,[14)] as follows

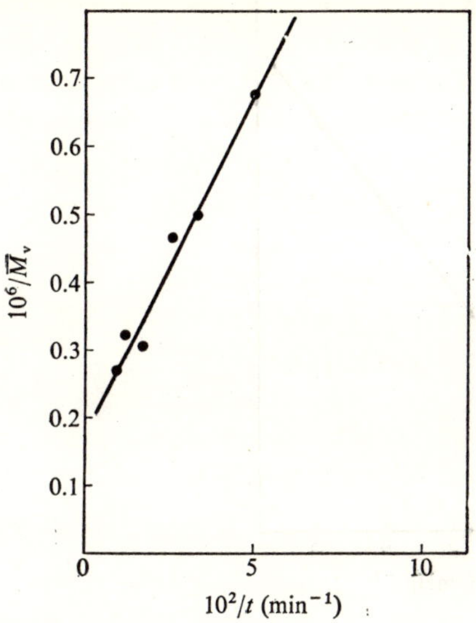

Fig. 5.20. Relationship between the reciprocals of molecular weight and polymerization time. ([A]/[Ti] = 2, 20°C.)

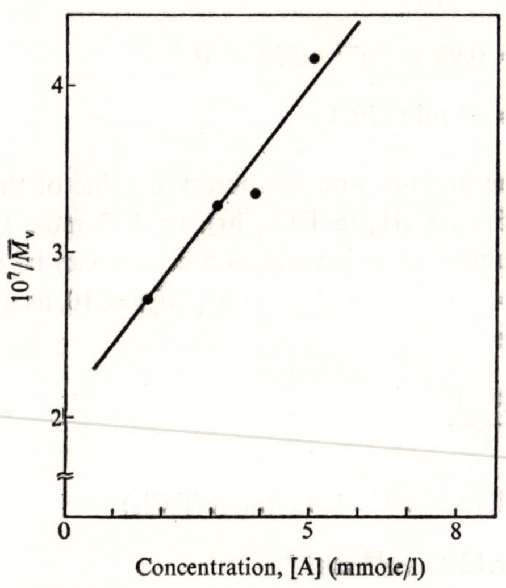

Fig. 5.21. Dependence of molecular weight on $AlEt_3$ concentration. (P_E = 40.5 cm Hg at 20°C.)

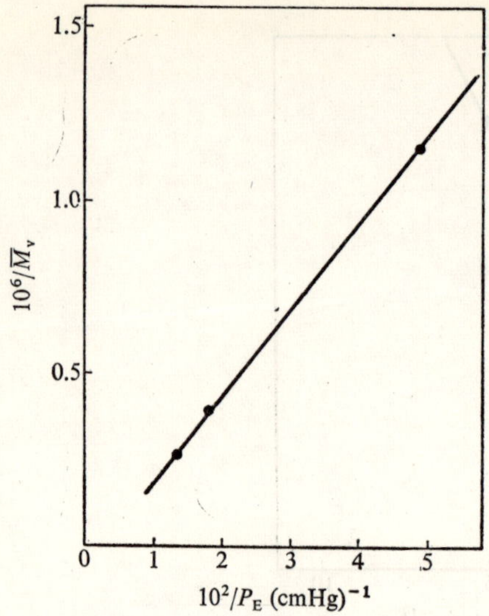

Fig. 5.22. Dependence of molecular weight on ethylene pressure. ([A]/[Ti] = 2, 20°C.)

$$\bar{L} = \frac{d(1/\overline{M}_v)}{d(1/t)} \overline{M}_v$$

$$= 0.98 \times 10^{-5} \times 35 \times 10^5$$

$$= 34 \text{ min } (20°\text{C}) \tag{5.13}$$

From this estimate of the lifetime, and the observed value of the stationary rate at 20°C (46.5 g-C_2H_4/(g-$TiCl_3$ hr), or 4.27 mole C_2H_4/(mole $TiCl_3$ min), the number of polymerization centres can be calculated. Assuming that $\bar{P}_n = \overline{M}_v/28\,Q$ and $Q\ (= \overline{M}_v/\overline{M}_n) = 10$, as quoted by Grieveson,[15] we have

$$[C^*] = \frac{R_\infty \bar{L}}{\bar{P}_n}$$

$$= 1.15 \times 10^{-2} \text{ mole/(mole-TiCl}_3)$$

$$= 1.12 \times 10^{14} \text{ cm}^{-2} \tag{5.14}$$

The specific surface area of the $TiCl_3$ was 40 m^2/g by the BET method. It may be noted that the value given by Eq. 5.14 for the number of polymerization centres at 20 °C is of the same order as the number of Ti atoms present on the catalyst surface.

Table 5.2 summarizes the effects of hydrogen on the polymerization rate and molecular weight. It can be seen that the value of the decay parameter, k_3, is not affected by the addition of hydrogen. The dependence of the stationary rate on hydrogen pressure can be expressed as

$$R_\infty^{H} = \frac{R_\infty^0}{1 + \alpha P_H^n} \tag{5.15}$$

TABLE 5.2 Ethylene polymerization in the presence of hydrogen[2)†]

T (°C)	P_H (cm Hg)	$[H_2] \times 10^4$ (mole/l)	P_E (cm Hg)	$\bar{R}_0 - R_\infty$ (g/(g-$TiCl_3$ hr))	R_∞	k_3 (hr^{-1})	$[\eta]$ (dl/g)	$\bar{M}_v \times 10^{-5}$
30	0	0	40	23	31.5	1.15	16.3	14.7
	6	2.89	40	20	20	1.18	10.7	8.33
	10	4.82	40	17	16.5	1.13	8.20	5.76
	20.5	9.88	40	12	13.5	1.18	6.70	4.36
40	0	0	40	23.1	41.0	2.12	15.3	13.6
	0.5	0.25	40	21.6	37.5	2.02	—	—
	1	0.51	40	21.0	35.0	1.92	13.5	11.5
	2	1.02	40	20.5	32.7	2.12	11.1	8.75
	3	1.53	40	18.0	30.0	2.14	—	—
	5	2.52	40	16.9	26.9	2.03	7.75	5.33
	10	5.06	40	16.0	20.6	2.27	7.15	4.77
	11	5.57	40	11.8	19.4	2.20	—	—
	20.5	10.4	40	9.4	18.5	2.06	4.54	2.54
	27	13.6	40	8.2	16.4	2.08	4.30	2.37
	10	5.06	30	10.0	15.0	1.80	—	—
	10	5.06	50	20.0	26.5	2.22	—	—
	10	5.06	60	25.0	31.3	2.39	—	—
50	0	0	40	30	48.3	2.45	13.0	10.9
	1	0.53	40	26	42.8	2.42	—	—
	5	2.64	40	22.5	30.0	2.30	6.15	3.88
	10	5.28	40	21	26.5	2.58	4.70	2.68
	15	7.92	40	19	20.5	2.40	4.10	2.22
	25	13.2	40	16.5	19.7	2.38	3.65	1.86

† $TiCl_3$(AA-grade, Stauffer Chem. Co.) 120 mg/150 ml toluene, [A]/[Ti] = 2.

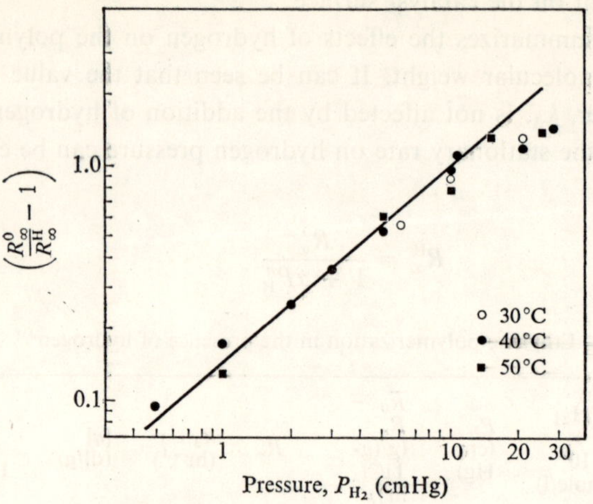

Fig. 5.23. Dependence of stationary rate on hydrogen pressure as given by Eq. 5.15. ([A]/[Ti] = 2.)

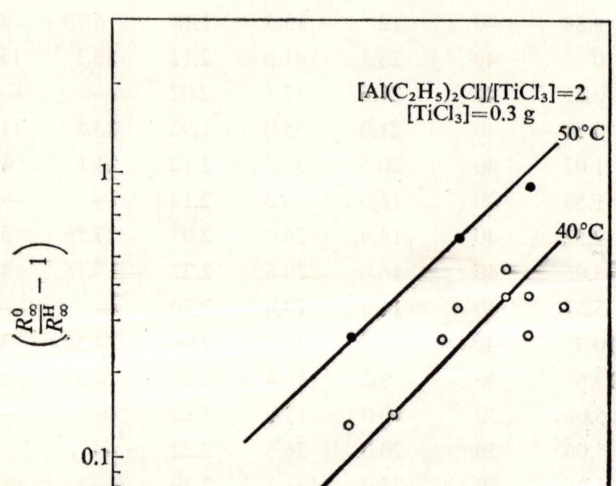

Fig. 5.24. Dependence of stationary rate on hydrogen pressure as given by Eq. 5.15, using $AlEt_2Cl$.

where the exponent, n, is 0.88 with $AlEt_3$ and 1.0 for $AlEt_2Cl$, as indicated by Fig. 5.23 and 5.24. Values for the constant α can be obtained from the intercepts in the figures, and are given in Table 5.3. The molecular weights of polymers produced in the presence of hydrogen are shown in Fig. 5.25, from which Eq. 5.16 can be derived.

TABLE 5.3 Constants α and β at various temperatures

Catalyst system	Constant $(mmHg)^{-1}$	Temp. (°C) 30	40	50	Activation energy (kcal/mole)
$TiCl_3$–$AlEt_3$	α	0.105	0.105	0.105	0
	β	0.128	0.215	0.299	14
$TiCl_3$–$AlEt_2Cl$	α		0.0145	0.035	8.6
	β		0.0107	0.022	12

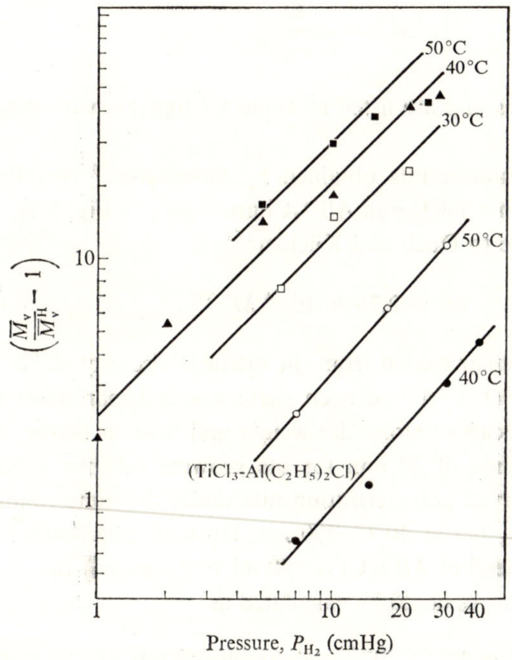

Fig. 5.25. Relationship between molecular weight and hydrogen pressure as given by Eq. 5.16.

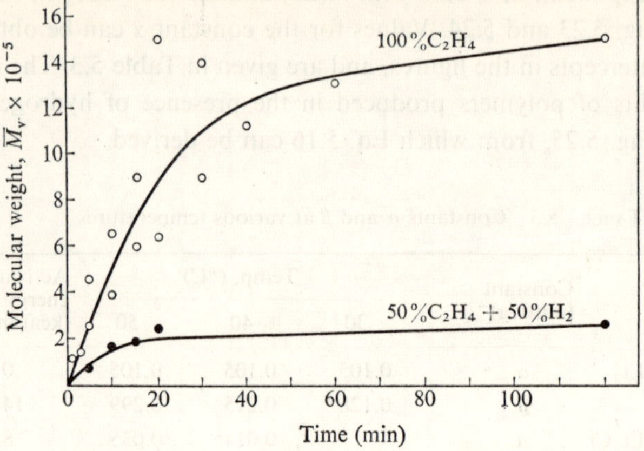

Fig. 5.26. Changes in molecular weight with polymerization time. ([γ-$TiCl_3$] = [$AlEt_2Cl$] = 5.25 mmole/l. P = 760 mmHg at 40°C.)

$$\overline{M}_v^H = \frac{\overline{M}_v}{1 + \beta P_H} \tag{5.16}$$

The observed values of β are listed in Table 5.3 together with those for α from Eq. 5.15.

The experimental results obtained by Grieveson[15] for ethylene polymerization with γ-$TiCl_3$ and $AlEt_2Cl$ are shown in Fig. 5.26, using the relationship due to Duch and Küchler[16]

$$[\eta] = 2.76 \times 10^{-4}\, \overline{M}_v^{0.78} \tag{5.17}$$

to obtain the molecular weight from the intrinsic viscosity at 120°C in decalin. In the absence of hydrogen there was a linear relationship between the reciprocals of molecular weight and time, as before. From the calculated lifetime of 30 min for the growing polymer chains at 40°C and the observed polymerization rate under the same conditions (4.8 l/(mmole-$TiCl_3$ hr) at 40°C, 590 mm Hg with 5.25 moles/l of γ-$TiCl_3$ and 5.25 moles/l of $AlEt_2Cl$ in 200 ml cyclohexane) the concentration of active centres could be calculated as

$$[C^*] = 1.5 \times 10^{-2} \text{ mole/(mole-}TiCl_3) \tag{5.18}$$

The corresponding value in the presence of 50% hydrogen was given as 6×10^{-3} mole/mole-$TiCl_3$ when $\bar{L} = 7$ min and the rate was 1.1 l/mmole-$TiCl_3$. The relationship between the degree of polymerization and the molecular weight, monomer concentration and hydrogen concentration was expressed as follows

$$\bar{P}_n = \frac{k_p[M][C^*]}{k_s[C^*] + k_M[M][C^*] + k_H[H_2][C^*]} \tag{5.19}$$

With $\bar{P}_n = K\bar{M}_v$, $K = 1/28Q$ and the dispersion ratio, $Q = \bar{M}_v/\bar{M}_n$, assumed equal to 10, Grieveson confirmed the applicability of Eq. 5.19 by plotting his results in the form shown in Fig. 5.27 and 5.28. Analysis of this data gave the following results (40 °C):

$$\begin{aligned} k_p &= 80 \text{ l/(mole sec)} \\ k_s &\simeq 3 \times 10^{-4} \text{ l/(mole sec)} \\ k_M &\simeq 1.1 \times 10^{-2} \text{ l/(mole sec)} \\ k_H &\simeq 3 \text{ l/(mole sec)} \end{aligned} \tag{5.20}$$

Grieveson accounted for these results in terms of three types of transfer reaction: (a) spontaneous termination by hydride transfer, (b) chain transfer by monomer, and (c) chain transfer by hydrogen

(a) $$\text{cat–CH}_2\text{–CH}_2 \xrightarrow{k_s} \text{cat–H} + \text{CH}_2\text{=CH}_2 \tag{5.21}$$

(b) $$\underset{\nwarrow \; \text{CH}_2\text{=CH}_2}{\text{cat–CH}_2\text{–CH}_2} \xrightarrow{k_M} \text{cat–CH}_2\text{–CH}_3 + \text{CH}_2\text{=CH} \tag{5.22}$$

(c) $$\underset{\nwarrow \; \text{H}_2}{\text{cat–CH}_2\text{–CH}_2} \xrightarrow{k_H} \text{cat–H} + \text{CH}_3\text{–CH} \tag{5.23}$$

As reported by Grieveson himself, the rate decrease due to hydrogen (Fig. 5.19) seems rather small compared with that found by other workers. Thus, a more rapid fall in the rate has been observed by Reich and Schindler,[17] whose results are shown in Fig. 5.29.

The results of Matsuda and Keii[2] are in good agreement with those of Grieveson except for the rate decrease due to hydrogen, where the data are similar to Schindler's observations.

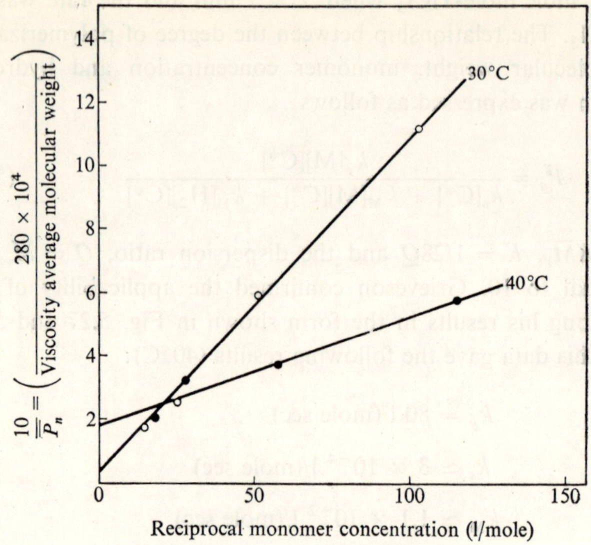

Fig. 5.27. Determination of the rate constant ratios k_s/k_p and k_M/k_p for the polymerization of ethylene at 30°C and 40°C (Eq. 5.19.)

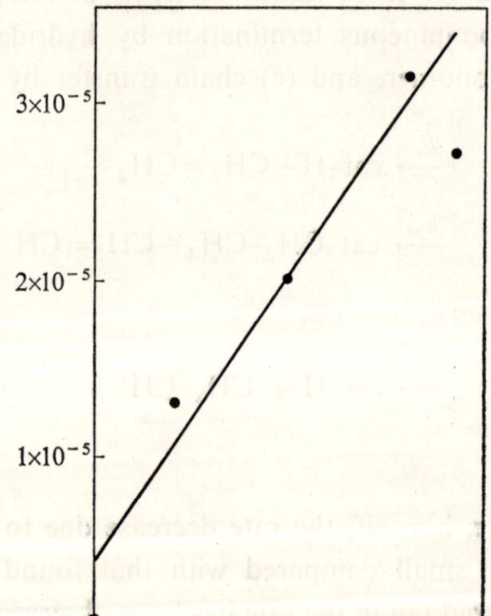

Fig. 5.28. Determination of the rate constant ratio k_H/k_p for polymerization of ethylene in the presence of hydrogen at 40°C.

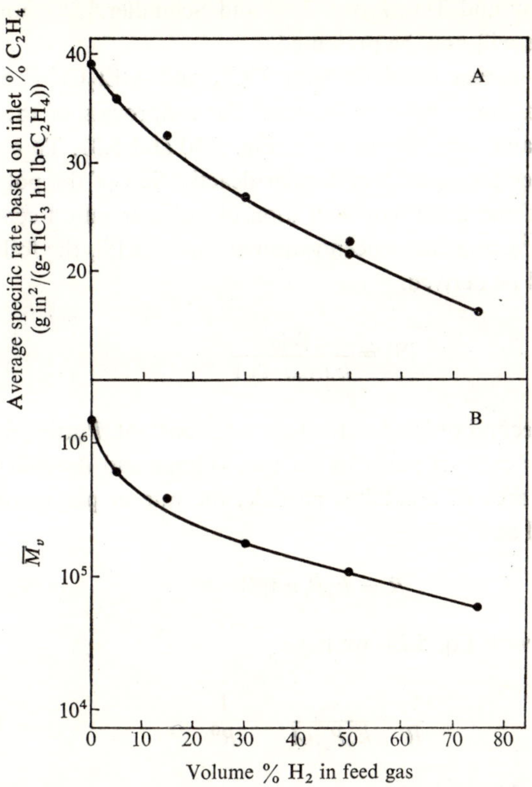

Fig. 5.29. A) Variation in specific polymerization rate with percentage of hydrogen in the feed gas, and B) change in molecular weight with percentage of hydrogen in the feed gas.[17)]

5.3 ETHYLENE POLYMERIZATION WITH ZIEGLER CATALYSTS

The polymerization of ethylene with catalyst systems based on $TiCl_4$ provides a typical case of Ziegler polymerization. The use of $TiCl_4$ with $AlEt_3$ or $AlEt_2Cl$ is the basis of industrial low-pressure processes for the manufacture of polyethylene, and has been widely studied, though less work has been carried out on the kinetics of these reactions. The kinetics of the $TiCl_4/AlEt_2Cl$ system have been investigated by Chirkov and Tsvetkova[18-23] and Schindler,[3,5,6] however, and these results will now be presented.

The kinetic curves obtained with $TiCl_4$ and $AlEt_2Cl$[3,5] can be represented by a linear relation between the reciprocals of the polymerization rate and time, as shown in Fig. 5.30 and 5.31. To account for this, Schindler proposed bimolecular deactivation of the polymerization centres. If the initial concentration of polymerization centres is $[S]_0$ (in mole/cm^2), and the concentration at time t is [S], the following expression could be derived,

$$[S] = \frac{[S]_0}{k_t[S]_0 t + 1} \tag{5.24}$$

where k_t is the second order deactivation rate constant. Putting n equal to the number of catalyst particles per unit volume and A equal to the average surface area of a catalyst particle, the rate of polymerization can be expressed as

$$R = k_p P_E nA[S]$$

Combining this with Eq. 5.24, we have

$$\frac{1}{R} = \frac{k_t}{k_p P_E nA} t + \frac{1}{R^0} \tag{5.25}$$

which is compatible with the observed relationship between rate and time, and which may be regarded as a form of Eq. 3.97. The quantity nA is proportional to the catalyst concentration, and the slope $(k_t/k_p P_E nA)$ should be proportional to the reciprocal of P_E and nA. Though this is the case with Schindler's results, the kinetic curves were observed only for the first 30 min. According to Chirkov *et al.*[18] the decrease in catalyst activity (in this case, 5.28 mmole/l $TiCl_4$, [A] = [Ti], in n-heptane) continues even after four hours at 30°C. From experimental results using [A]/[Ti] = 0.5 and 4.0 at 30°C, Chirkov *et al.*[19] concluded that the higher the catalyst concentration, the shorter the time required to reach the stable state. They considered that catalyst deactivation was due to excess $AlEt_2Cl$, based on the observation that deactivation could be minimized by dilution of the catalyst system with solvent.[20] However, this is compatible with the theory of Schindler since the effect may arise from the nA term in Eq. 5.25. It appears that

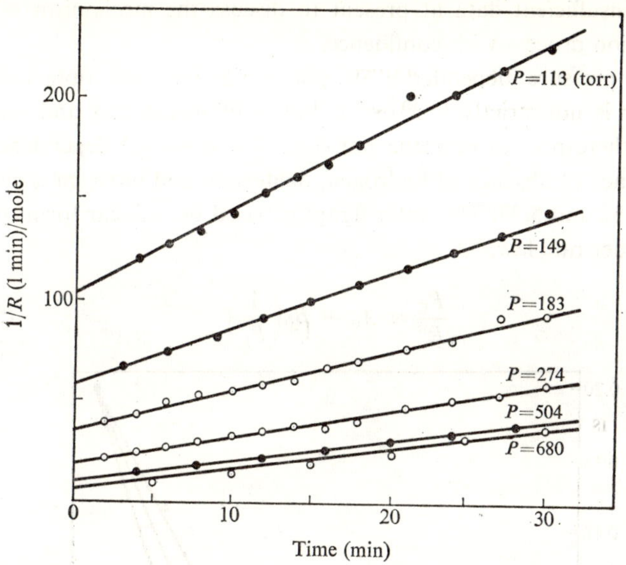

Fig. 5.30. Time dependence of the rate of polymerization using preset ethylene pressures in the presence (○) and absence (●) of nitrogen.

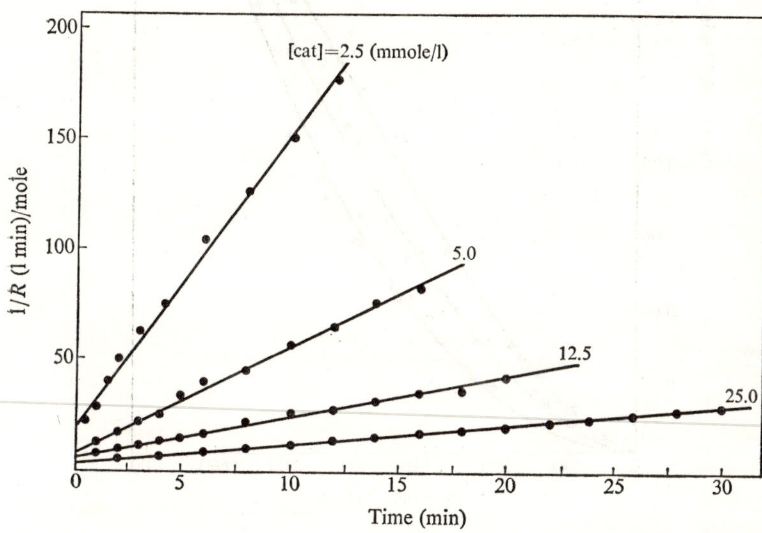

Fig. 5.31. Effect of catalyst concentration ($[AlEt_2Cl]/[TiCl_4] = 1$) on the kinetic curves.

there is insufficient data at present to discuss the mechanism of the deactivation process with confidence.

The pressure dependence of polymerization rate observed by Schindler is not strictly first order, but is of mixed first and second order with respect to ethylene pressure. The observed dependence in the presence or absence of hydrogen, deuterium and nitrogen is shown in Fig. 5.32 and 5.33. The latter figure is based on a linear form of the mixed order rate law,

$$\frac{P_E}{R^0} = A_0 + B_0\left(\frac{1}{P_E}\right) \tag{5.26}$$

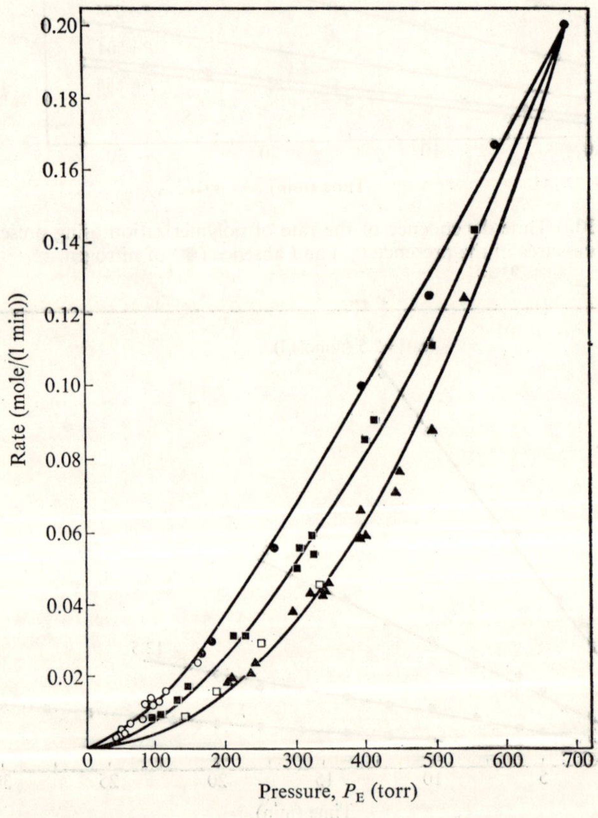

Fig. 5.32. Pressure dependence of ethylene polymerization rate in *n*-heptane at 40°C. ([$TiCl_4$] = 3.0 mmole/l, [$AlEt_2Cl$] = 4.5 mmole/l, catalyst system was aged for 30 min. Preset ethylene pressures in the presence of N_2 (■), H_2 (▲), D_2 (□), and in the absence of diluent gases (●). Total pressure was 770 torr. Data obtained at preset feed rate is shown by ○.)

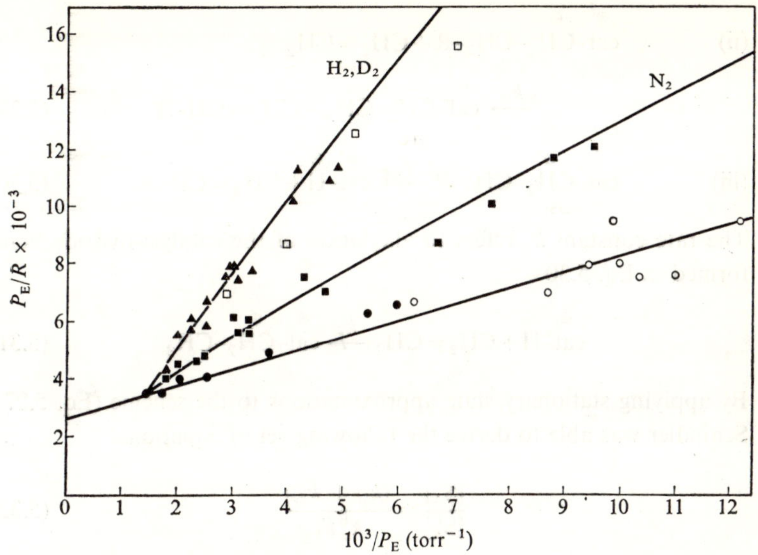

Fig. 5.33. Pressure dependence of the polymerization rate as given by Eq. 5.26. (Symbols as in Fig. 5.32.)

which is of the same functional form as Eq. 5.1. The figures clearly show the retardation effect of the diluent gases. The results were explained by Schindler in terms of the following kinetic scheme,

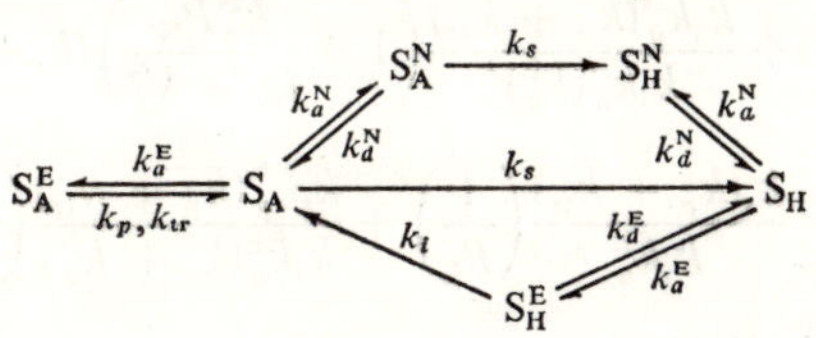

(5.27)

where subscripts A and H indicate sites with active alkyl and hydride bonds, and superscripts E, N and H refer to adsorbed molecules of ethylene, nitrogen and hydrogen respectively. The rate constants k_a, k_d, k_p, k_{tr} and k_s refer to adsorption, desorption, propagation (i), chain transfer by monomer (ii) and spontaneous disproportionation by hydrogen transfer from the β-carbon of the chain to the catalyst metal atom (iii).

(i) $$\overset{\oplus}{\text{cat}}\text{-}\overset{\ominus}{\text{CH}_2}\text{-CH}_2\text{-(CH}_2\text{-CH}_2)_n\text{-R} + \text{CH}_2{=}\text{CH}_2$$

$$\xrightarrow{k_p} \overset{\oplus}{\text{cat}}\text{-}\overset{\ominus}{\text{CH}_2}\text{-CH}_2\text{-(CH}_2\text{-CH}_2)_{n+1}\text{-R} \qquad (5.28)$$

(ii) $$\overset{\oplus}{\text{cat}}\text{–}\overset{\ominus}{\text{CH}_2}\text{–CH}_2\text{–R}+\text{CH}_2=\text{CH}_2 \xrightarrow{k_{tr}} \overset{\oplus}{\text{cat}}\text{–}\overset{\ominus}{\text{CH}_2}\text{–CH}_3+\text{CH}_2=\text{CH–R} \quad (5.29)$$

(iii) $$\overset{\oplus}{\text{cat}}\text{–}\overset{\ominus}{\text{CH}_2}\text{–CH}_2\text{–R} \xrightarrow{k_s} \overset{\oplus}{\text{cat}}\text{–}\overset{\ominus}{\text{H}}+\text{CH}_2=\text{CH–R} \quad (5.30)$$

The rate constant k_i refers to alkylation of the catalyst-hydride bond formed in Eq. 5.30

$$\overset{\oplus}{\text{cat}}\text{–}\overset{\ominus}{\text{H}}+\text{CH}_2=\text{CH}_2 \xrightarrow{k_i} \overset{\oplus}{\text{cat}}\text{–}\overset{\ominus}{\text{CH}_2}\text{–CH}_3 \quad (5.31)$$

By applying stationary state approximations to the scheme (Eq. 5.27), Schindler was able to derive the following set of equations.

$$\frac{[\text{S}_\text{A}]}{[\text{S}_\text{A}^\text{E}]} = \frac{(k_p + k_{tr})}{k_a^\text{E} P_\text{E}} \quad (5.32)$$

$$\frac{[\text{S}_\text{A}^\text{N}]}{[\text{S}_\text{A}^\text{E}]} = \frac{k_a^\text{N}(k_p + k_{tr})}{k_a^\text{E}(k_s + k_d^\text{N})}\left(\frac{P_\text{N}}{P_\text{E}}\right) \quad (5.33)$$

$$\frac{[\text{S}_\text{H}^\text{N}]}{[\text{S}_\text{A}^\text{E}]} = \frac{k_s k_a^\text{N}(k_p + k_{tr})}{k_a^\text{E} k_d^\text{N}(k_s + k_d^\text{N})}\left(\frac{P_\text{N}}{P_\text{E}}\right) + \frac{k_s k_a^\text{N}(k_p + k_{tr})}{k_i k_d^\text{N}(k_a^\text{E})^2}\left(1 + \frac{k_a^\text{N} P_\text{N}}{k_s + k_d^\text{N}}\right)(k_i + k_d^\text{E})\left(\frac{P_\text{N}}{P_\text{E}^2}\right) \quad (5.34)$$

$$\frac{[\text{S}_\text{H}^\text{E}]}{[\text{S}_\text{A}^\text{E}]} = \frac{k_s(k_p + k_{tr})}{k_i k_a^\text{E}}\left(\frac{1}{P_\text{E}}\right) + \frac{k_s k_a^\text{N}(k_p + k_{tr})}{k_i k_a^\text{E}(k_s + k_d^\text{N})}\left(\frac{P_\text{N}}{P_\text{E}}\right) \quad (5.35)$$

$$\frac{[\text{S}_\text{H}]}{[\text{S}_\text{A}^\text{E}]} = \frac{k_s(k_p + k_{tr})(k_i + k_d^\text{E})}{k_i(k_a^\text{E})^2}\left(\frac{1}{P_\text{E}^2}\right) + \frac{k_s k_a^\text{N}(k_p + k_{tr})(k_i + k_d^\text{E})}{k_i(k_a^\text{E})^2(k_s + k_d^\text{N})}\left(\frac{P_\text{N}}{P_\text{E}^2}\right) \quad (5.36)$$

The rate of polymerization is given by

$$R = k_p[\text{S}_\text{A}^\text{E}] \quad (5.37)$$

and its dependence on the total concentration of catalyst sites, $[\text{S}_t]$, and the partial gas pressures, P_E and P_N, can be obtained from the following equation.

$$\frac{k_p[S_t]}{R} = \frac{[S_t]}{[S_A^E]} = \frac{[S_A]}{[S_A^E]} + \frac{[S_A^N]}{[S_A^E]} + \frac{[S_H^N]}{[S_A^E]} + \frac{[S_H^E]}{[S_A^E]} + \frac{[S_H]}{[S_A^E]} + 1 \tag{5.38}$$

Assuming $k_p \gg k_{tr}$, we then have

$$\frac{P_E}{R} = A_0 + B_0\left(\frac{1}{P_E}\right) + B_0 C_N\left(\frac{P_N}{P_E}\right) + A_0 C_N P_N + B_0 C_N'(1 + C_N P_N)\left(\frac{P_N}{P_E}\right) + C_N\left(\frac{k_s}{k_a^E k_d^N [S_t]}\right) P_N + \left(\frac{1}{k_p[S_t]}\right) P_E \tag{5.39}$$

where

$$A_0 = \left(\frac{1}{k_a^E[S_t]}\right)\left(\frac{k_i + k_s}{k_i}\right)$$

$$B_0 = \left(\frac{k_s}{k_a^E}\right)\left(\frac{1}{k_a^E[S_t]}\right)\left(\frac{k_i + k_d^E}{k_i}\right)$$

$$C_N = \frac{k_a^N}{k_s + k_d^N} \tag{5.40}$$

$$C_N' = \frac{k_a^N}{k_d^N}$$

Schindler assumed that the adsorption of ethylene is the rate determining step, and was therefore able to neglect the last term in Eq. 5.38 and 5.39. In the case where there are no diluent gases, Eq. 5.39 reduces to Eq. 5.26. In the presence of nitrogen, Schindler considered that the penultimate term of Eq. 5.39 could be neglected and that C_N was approximately equal to C_N', on the basis of experimental observations. This gave rise to Eq. 5.41

$$\frac{P_E}{R^N} = \frac{P_E}{R^0} + 2B_0 C_N\left(\frac{P_N}{P_E}\right) + A_0 C_N P_N \tag{5.41}$$

A similar treatment could be applied in considering the presence of hydrogen in the system, though an additional path ($S_A^H \rightarrow S_H$) must then be introduced, based on the following reaction

$$\overset{\oplus}{\text{cat}}\text{-}\overset{\ominus}{\text{CH}_2}\text{-CH}_2\text{-R} + \text{H}_2 \xrightarrow{k_H} \overset{\oplus}{\text{cat}}\text{-}\overset{\ominus}{\text{H}} + \text{CH}_3\text{-CH}_2\text{-R} \tag{5.42}$$

The pressure dependence of the rate in the presence of hydrogen could now be expressed as

$$\frac{P_E}{R^H} = \frac{P_E}{R^0} + B_0 C_H \left(\frac{2 + k_H}{k_\backslash}\right)\left(\frac{P_H}{P_E}\right) + A_0 C_H P_H \tag{5.43}$$

where $C_H = k_a^H/k_d^H$. Eq. 5.26, 5.41 and 5.43 were tested experimentally as shown in Fig. 5.32 and 5.33, subject to the important restriction that the sum of the various partial pressures remained constant. In this case, the theoretical equations (e.g. Eq. 5.41) could be rewritten in the following form.

$$\frac{P_E}{R^N} = \frac{P_E}{R^0} + 2B_0 C_N \left(\frac{P^0 - P_E}{P_E}\right) + A_0 C_N (P^0 - P_E) \tag{5.44}$$

where P^0 represents the overall pressure, which in this case is equal to the sum of the partial pressures of ethylene and nitrogen. This form can account for the experimental linearity between P_E/R^N and P_E provided that the last term could be neglected. This may be valid, since the values of C_N or C'_N, corresponding to the adsorption equilibrium constant, are small. However, if this last term is neglected, e.g. in Eq. 5.41, then the second term must also be negligible, and the equation reduces to that expected in the absence of nitrogen.

$$\begin{aligned} \frac{P_E}{R^N} &= A_0(1 + C_N P_N) + B_0 \left(\frac{1 + 2C_N P_N}{P_E}\right) \\ &\simeq A_0 + B_0 \left(\frac{1}{P_E}\right) \\ &= \frac{P_E}{R^0} \end{aligned} \tag{5.45}$$

It therefore seems to the present author that the experimental linearity found by Schindler must be accounted for in some other way.

According to Chirkov *et al.*[22] the rate of polymerization is directly proportional to monomer concentration up to 1.4 atm at 30°C ([A]/[Ti] = 1.0 with [$TiCl_4$] = 10 g/l). Schnecko *et al.*[9] reported that the rate dependence on $AlEt_2Cl$ concentration was a Langmuir type, whereas with $AlEt_3$, it showed a maximum, as indicated also by Fukui *et al.*[24]

The preceding discussion of the experimental behavior of Ziegler catalysts based on $TiCl_4$ has shown that greater complexities are to be found than with the simpler Ziegler-Natta systems. This is probably the case because of the complex reactions involved in catalyst formation, and a full understanding of Ziegler catalysts will probably not be attained until Ziegler-Natta catalyst behavior can be accounted for quite thoroughly.

In conclusion, a simple model will be presented for the behaviour of the polymerization rate in the presence of diluent gases. If adsorption control is assumed, as it was in Schindler's model, the state S_A^E can be neglected, and all the remaining states will be in equilibrium. The coverage of the respective sites may then be expressed as

$$\theta_N = \frac{K_N P_N}{1 + K_N P_N + K_H P_H} \tag{5.46a}$$

$$\theta_H = \frac{K_H P_H}{1 + K_N P_N + K_H P_H} \tag{5.46b}$$

and the rate of polymerization will be

$$\begin{aligned} R &= k_p P_E \theta_A (1 - \theta_N - \theta_H) \\ &= \frac{k_p P_E \theta_A}{1 + K_N P_N + K_H P_H} \end{aligned} \tag{5.46c}$$

Because of the processes shown in Eq. 5.30 and 5.31, the state S_H can be assumed to be in equilibrium even in the absence of diluent gases

$$k_s \theta_A = k_i P_E \theta_H \tag{5.46d}$$

and so, denoting the sum of θ_A and θ_H by θ_A^0, we have

$$\theta_A = \frac{\theta_A^0}{1 + \dfrac{k_s}{k_i P_E}} \tag{5.46e}$$

Eq. 5.46*c* can therefore be rewritten as

$$R = \frac{k_p P_E \theta_A^0}{\left(1 + \dfrac{k_s}{k_i P_E}\right)(1 + K_N P_N + K_H P_H)} \tag{5.46f}$$

from which we obtain

$$\frac{P_E}{R} = \frac{1}{k_p\theta_A^0}\left[1 + \frac{k_s}{k_i}\left(\frac{1}{P_E}\right)\right][1 + K_N P_N + K_H P_H]$$

$$= A_0' + B_0'\left(\frac{1}{P_E}\right) + B_0' K_N\left(\frac{P_N}{P_E}\right) + B_0' K_H\left(\frac{P_H}{P_E}\right)$$

$$+ A_0' K_N P_N + A_0' K_H P_H \qquad (5.46g)$$

This equation corresponds to Eq. 5.39, and is equally valid in terms of the experimental observations.

References

1. M.N. Berger and B.M. Grieveson, *Makromol. Chem.*, **83**, 80 (1965).
2. T. Matsuda, T. Keii, A. Kojima and T. Ishida, *Kogyo Kagaku Zasshi* (Japanese), **71**, 1124, 1128 (1968); T. Matsuda, *Dr. Eng. Thesis, Tokyo Inst. Technology*, 1968.
3. A. Schindler, *J. Polymer Sci.*, **C4**, 81 (1963).
4. W.E. Smith and R.G. Zelmer, *J. Polymer Sci.*, **A1**, 2587 (1963).
5. A.Schindler, *Makromol. Chem.*, **73**, 94 (1964).
6. A. Schindler, unpublished paper; L. Reich and A. Schindler, *Polymerization by Organometallic Compounds*, p. 320, Interscience, 1966.
7. H. Schnecko, M. Reinmoeller, K. Weirauch, W. Lintz and W. Kern, *Makromol. Chem.*, **69**, 105 (1963).
8. H. Schnecko, M. Reinmoeller, K. Weirauch and W. Kern, *J. Polymer Sci.*, **C4**, 71 (1963).
9. H. Schnecko, M. Reinmoeller, K. Weirauch, V. Bednjagin and W. Kern, *Makromol. Chem.*, **73**, 154 (1964).
10. L.M. Lanovskaya, N.V. Makletsova, A.R. Gantmakher and S.S. Medvedev, *Vysokomolekul. Soedin.*, 7, 741 (1965).
11. *ibid.*, **7**, 747 (1965).
12. K. Fukui, T. Kagiya, S. Machi, T Shimizu and S. Yuasa, *Bull. Chem. Soc. Japan*, **35**, 303, 306 (1962).
13. L.H. Tung, *J. Polymer Sci.*, **24**, 333 (1957).
14. G. Natta and I. Pasquon, *Advan. Catalysis*, **11**, 1 (1959).
15. B.M. Grieveson, *Makromol. Chem.*, **84**, 93 (1965).
16. E. Duch and L. Küchler, *Z. Elektrochem.*, **60**, 218 (1956).
17. L. Reich and A. Schindler, *Polymer Rev.*, **12**, 326 (1966).
18. E. Varadi, V.I. Tsvetkova and N.M. Chirkov, *Dokl. Akad. Nauk SSSR*, **152** (1963).
19. I.M. Meshkova, V.I. Tsvetkova and N.M. Chirkov, *Izv. Akad. Nauk SSSR, Otd. Khim. Nauk.*, **1963**, 515.
20. *ibid.*, **1964**, 386.
21. I.M. Meshkova, S.J. Kuo, V.I. Tsvetkova and N.M. Chirkov, *Vysokomolekul. Soedin.*, **5**, 10 (1963).
22. J. Varady, V.I. Tsvetkova and N.M. Chirkov, *Magy. Kem. Folyoirat.*, **70**, 264

(1964).
23. *ibid.*, **71**, 136 (1965).
24. S. Kodama, T. Kagiya, S. Machi, T. Shimizu, S. Yuasa and K. Fukui, *J. Appl. Polymer Sci.*, **3**, 20 (1960).

Chapter 6

Effects of electron-pair donors and surface heterogeneity

6.1 INTERACTION OF ELECTRON-PAIR DONORS WITH CATALYST AND COCATALYST

Many reports and technical patents have been published on the influence of additives during Ziegler-Natta polymerization. In many ways, the most interesting additives are electron-pair donor molecules, which can affect both the activity and the stereospecificity of the catalysts.

The effects of electron-pair donors can generally be understood on the basis of interactions with the cocatalyst and/or the catalyst in Ziegler-Natta systems. According to Hoeg *et al.*,[1)] an electron-pair donor may react with an organometal to form an addition compound, as shown in Eq. 6.1 for the case of pyridine and $AlEt_3$.

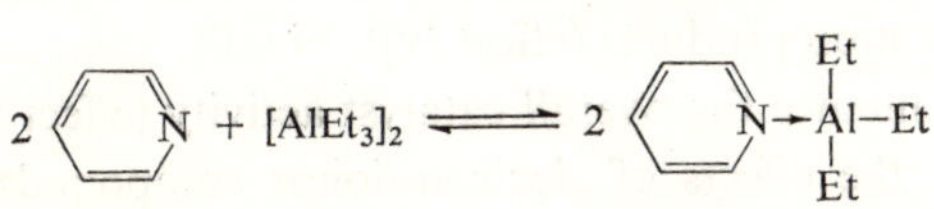

(6.1)

The equilibrium constant for this reaction has been found to be 25 for $AlEt_3/Et_3N$,[2)] which means that the formation of the addition compound is essentially irreversible. The reactivity of electron donor molecules with organoaluminium compounds decreases in the order N-containing > 0– > S–. The effects of a reaction such as Eq. 6.1 are (i) to decrease the effective concentration of cocatalyst present, and (ii) the influence of the products of the reaction, e.g. poisoning of the catalyst surface sites by adsorption of the product.

A third effect, (iii), has been noted in the reaction between the electron-donor molecule HPT (hexamethylphosphoric amide) and $AlEtCl_2$.

$$2AlEtCl_2 + HPT \rightarrow AlEt_2Cl + AlCl_3 \cdot HPT \quad (6.2)$$

This may be regarded as a simple disproportionation reaction, 2 $AlEtCl_2 \rightarrow AlEt_2Cl + AlCl_3$, occurring in the presence of HPT. $AlEtCl_2$ has no activity in propylene polymerization, but one of the products, $AlEt_2Cl$, is well known as an active and highly stereospecific cocatalyst, and may therefore influence the course of the polymerization.

$TiCl_3$ is a strong Lewis acid (the acidity has been measured, using the pyridine chemisorption method, by Tanabe *et al.*[4]) and Kanetaka[5])) and electron-donor molecules such as amines are readily adsorbed on the surface. This gives rise to effect (iv), a reduction in catalyst activity due to the reduced availability of surface sites for polymerization. In some cases, quite complex reactions may occur, involving the electron donor, catalyst and solvent molecules. The following reaction[7]) provides an example

$$2TiCl_3 + O_2 + C_7H_8(\text{solvent}) \rightarrow TiOCl_2 + TiCl_4 + C_7H_7OH \quad (6.3)$$

$$C_7H_7OH + TiCl_4 \rightarrow (C_7H_7O)TiCl_3 \text{ or } (C_7H_7O)TiCl_2 \quad (6.3a)$$

or

$$2TiCl_3 + \tfrac{1}{2}O_2 \rightarrow TiOCl_2 + TiCl_4 \quad (6.3b)$$

In this case, the formation of a new catalyst, $TiCl_4$, changes the nature of the polymerization products (effect (v)). Finally, some additives have been found to enhance the overall catalyst activity (effect (vi)).

In general, the effects of electron-donor compounds on Ziegler-Natta polymerization can be described in terms of the above six effects. It should be noted that the order of addition of catalyst, cocatalyst and electron-donor will affect the results.

6.2 INFLUENCE OF ELECTRON DONORS ON THE $TiCl_3/AlEt_3$ SYSTEM

Some effects arising from the presence of electron donor molecules in the $TiCl_3/AlEt_3$ catalyst system include a decrease in polymerization activity, and increase in molecular weight[8,9] and isotacticity[10,11] of the polymer produced. Vessely *et al.*[8] found that electron donor additives such as alcohols, acetone or amines formed complexes with $AlEt_3$, whereas COS, CS_2 and similar compounds acted as poisons for the $TiCl_3$ catalyst.

In this section the effects of $SeOCl_2$ and O_2 will be considered in detail with particular reference to their kinetic effects. Soga *et al.*[10] investigated the influence of $SeOCl_2$, using different orders of addition. In case A the order was ($SeOCl_2$ + $AlEt_3$) + $TiCl_3$ and in case B ($SeOCl_2$ + $TiCl_3$) + $AlEt_3$. The dependence of the stationary rate on the amount of $SeOCl_2$ found for case A is shown in Fig. 6.1, and it can be seen that polymerization can no longer proceed when the amount of $SeOCl_2$ added reaches 1.6 mmole, corresponding to half the total amount of $AlEt_3$ present. We may therefore assume the ir-

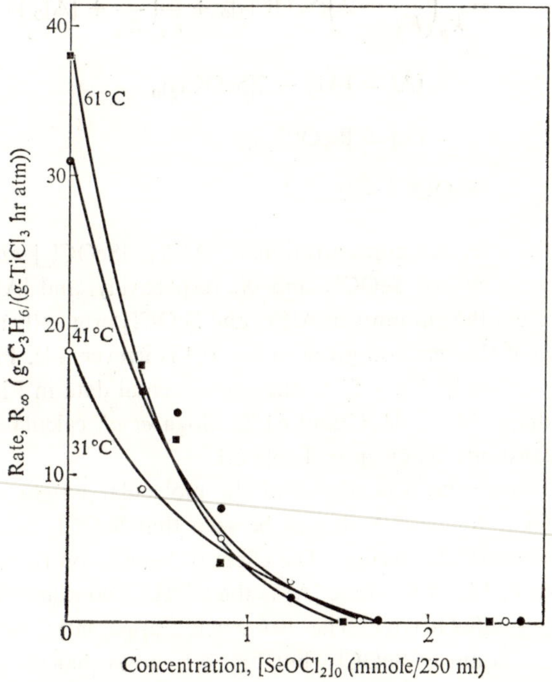

Fig. 6.1. Dependence of the stationary rate of propylene polymerization on $SeOCl_2$ concentration. ($[TiCl_3]$ = 3.2 mmole/250 ml toluene. [A]/[Ti] = 1.0.)

reversible formation of a complex X

$$Al_2Et_6 + SeOCl_2 \rightarrow X \tag{6.4}$$

It was further assumed that X was inactive, but was adsorbed on the $TiCl_3$ surface (i.e. effects (i) and (ii) both occur).

The stationary rate of polymerization under these conditions can be expressed as

$$R_\infty = \frac{k_p K_A[A]}{1 + K_A[A] + K_S[SeOCl_2] + K_X[X]} \tag{6.5}$$

$$= \frac{k_p K_A([A]_0 - 2[SeOCl_2]_0)}{1 + K_A([A]_0 - 2[SeOCl_2]_0) + K_X[SeOCl_2]_0} \tag{6.5a}$$

or

$$\frac{[A]_0 - 2[SeOCl_2]_0}{R_\infty} = \frac{1}{k_p}\left(\frac{K_X}{K_A} - 2\right)[SeOCl_2]_0 + \frac{1}{k_p}\left(\frac{1}{K_A} + [A]_0\right) \tag{6.5b}$$

with

$$[A] = [A]_0 - 2[SeOCl_2]_0$$

$$[X] = [SeOCl_2]_0 \tag{6.5c}$$

$$[SeOCl_2] = 0$$

where [A] is the effective concentration of $AlEt_3$, $[SeOCl_2]$ and [X] are the concentrations of $SeOCl_2$ and X, respectively, and $[A]_0$ and $[SeOCl_2]_0$ represent the amounts of $AlEt_3$ and $SeOCl_2$ originally added. Eq. 6.5*c* follows if the reaction given in Eq. 6.4 is irreversible. Fig. 6.2 shows the application of Eq. 6.5*b* to the experimental data in Fig. 6.1. From the results at 31 °C, 41 °C and 61 °C, Soga *et al.* calculated the values of the constants, as given in Table 6.1.

Table 6.2 shows the dependence of the molecular weight on the amount of $SeOCl_2$ for case A. It can be seen that $SeOCl_2$ exerts no influence on the molecular weight. The effect of $SeOCl_2$ on isotacticity has been shown in Fig. 4.1, where [A] is the effective concentration of $AlEt_3$ (= $[A]_0 - 2[SeOCl_2]_0$). The isotacticity apparently increases with increasing amount of $SeOCl_2$, but Fig. 4.1 shows that this is due only to the reduction in effective concentration of $AlEt_3$,[12] and that actual $AlEt_3$ concentrations (in the absence of $SeOCl_2$) equal to the

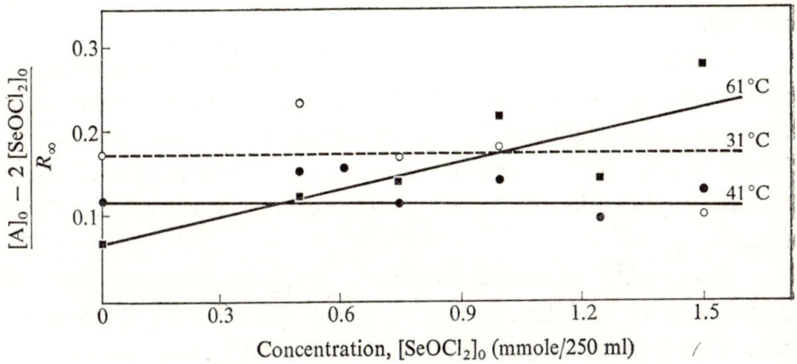

Fig. 6.2. Applicability of Eq. 6.5 to the data shown in Fig. 6.1.

TABLE 6.1 Values of rate constants and heats of adsorption

Temp. (°C)	k (g-C_3H_6/(g-$TiCl_3$ hr atm))	K_A† (l/mmole)	E_A (kcal/mole)	K_X (l/mmole)	E_X (kcal/mole)
31	18	3.70	12	4.5∼6.5	2∼3
41	32	2.04	12	4.1	2∼3
61	38	0.65	12	3.2	2∼3

† From propylene polymerization with $TiCl_3$ and $AlEt_3$.

TABLE 6.2 Dependence of molecular weight on amount of $SeOCl_2$ added at 41 °C

$[SeOCl_2]_0$ (mmole/250 ml)	$\bar{M}_v \times 10^{-5}$
0	4.7
0.50	5.0
0.75	5.2
1.25	5.0

effective concentrations (in the presence of $SeOCl_2$) have the same effect on isotacticity. Thus the effects of $SeOCl_2$ are limited to (i) and (ii).

However, in case B the role of the $SeOCl_2$ is different. Stationary rates in this case were very small compared with case A, though neither the molecular weight nor the isotacticity of the polymer was affected. These results (summarized in Table 6.3) could be attributed simply to strong adsorption of $SeOCl_2$ on the catalyst surface, i.e. to effect (iv).

Doi *et al.*[11] examined the effects of O_2 on propylene polymerization with $TiCl_3/AlEt_3$. In case A, $(O_2 + AlEt_3) + TiCl_3$, the results were similar to those found for case A with $SeOCl_2$. Doi *et al.* assumed that an irreversible reaction of the type

$$Al_2Et_6 + \frac{3}{2}O_2 \rightarrow X \tag{6.6}$$

took place, and derived a rate law in the same way as for $SeOCl_2$.

$$R_\infty = \frac{k_p K_A([A]_0 - \frac{4}{3}[O_2]_0}{1 + K_A([A]_0 - \frac{4}{3}[O_2]_0) + \frac{2}{3}K_X[O_2]_0} \tag{6.7}$$

Fig. 6.3 shows a plot of $(3[A]_0 - 4[O_2]_0)/R$ against $[O_2]_0$, based on the linear form of Eq. 6.7. The temperature dependence of the rate k_p and the adsorption constants K_A and K_X are shown in Fig. 6.4. The results obtained in case A are summarized in Table 6.4.

The increase in isotacticity with amount of O_2 (Fig. 6.5) could be explained in terms of the decrease in effective concentration of $AlEt_3$, in the same way as for $SeOCl_2$. Doi *et al.*[11] supposed that X might represent $(EtO)_3Al_2Et_3$ which may be produced as an intermediate during $Al_2(OEt)_6$ formation. Ziegler *et al.*[13] confirmed the formation

TABLE 6.3 Results of polymerization (Case B) at 41 °C

$[SeOCl_2]_0$ (mmole/250 ml)	R_∞ (g-C_3H_6/(g-$TiCl_3$ hr atm))	$\overline{M}_v \times 10^{-5}$	Isotacticity (%)
0	32	4.7	81
0.20	12	4.6	82
0.40	4.0	5.1	81
0.80	0	4.9	—

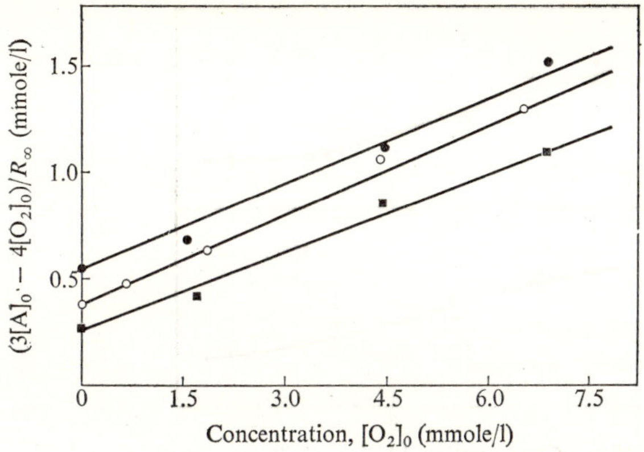

Fig. 6.3. Applicability of Eq. 6.7. to the data of Doi *et al.*[11)]

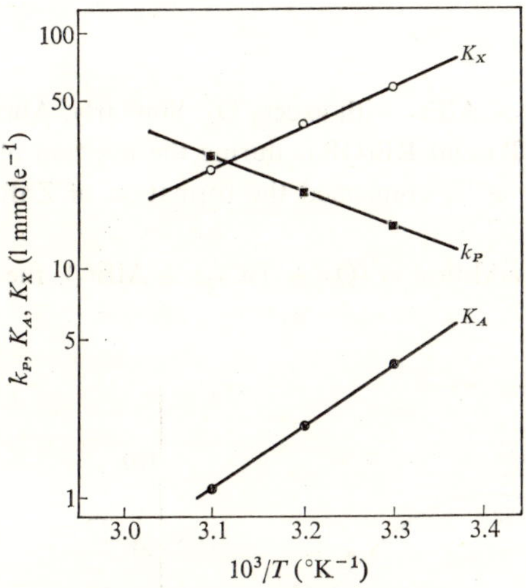

Fig. 6.4. Temperature dependence of the rate constants of Eq. 6.7.

TABLE 6.4 Rate constants of rate law, Eq. 6.7

Temp. (°C)	k_p (g-C_3H_6/ (g-$TiCl_3$ hr atm))	E_p (kcal/ mole)	K_A (l/mmole)	E_A (kcal/ mole)	K_X (l/mmole)	E_X (kcal/ mole)
30	15	7.1	3.7	12	59	8.0
40	22	7.1	2.0	12	41	8.0
50	30	7.1	1.1	12	26	8.0

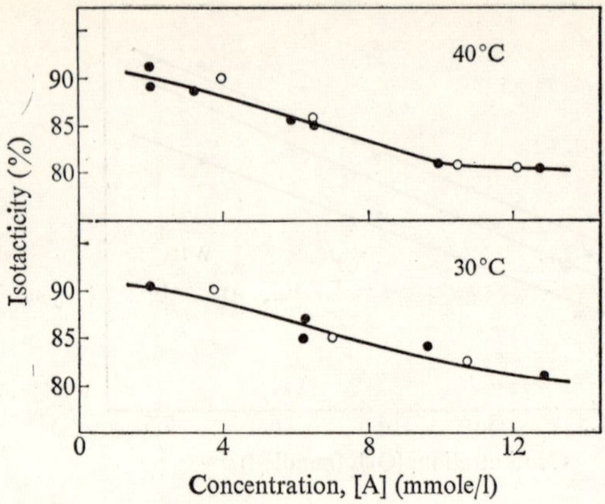

Fig. 6.5. Relationship between isotacticity and $AlEt_3$ concentration. (Data obtained with $TiCl_3/AlEt_3$ is shown as ● and that obtained with $TiCl_3/(AlEt_3 + O_2)$ by ○.)

of $Al_2(OEt)_6$ by reaction of $AlEt_3$ with excess O_2. Similarly, Abraham *et al.*[14] obtained $R_2B(OR)$ and $RB(OR)_2$ during the reaction of BR_3 with O_2, and Tsuruta *et al.*[15] confirmed the formation of ZnR(OR) and $Zn(OR)_2$.

In case B (order of addition is $(O_2 + TiCl_3) + AlEt_3$), the most

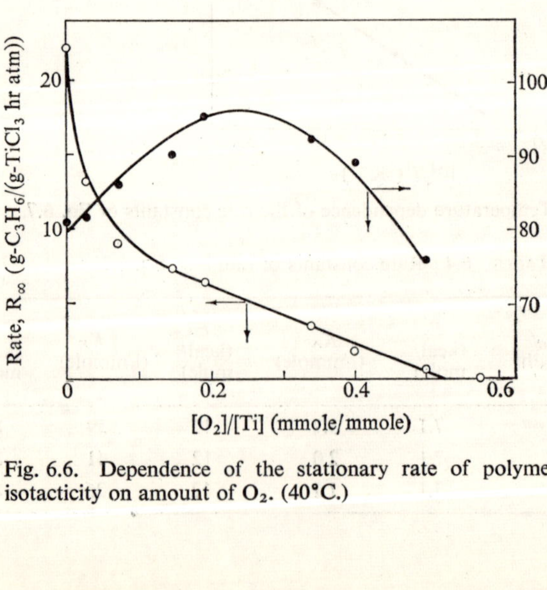

Fig. 6.6. Dependence of the stationary rate of polymerization and the isotacticity on amount of O_2. (40°C.)

striking result is the existence of an optimum amount of O_2 for producing an increase in isotacticity, as shown in Fig. 6.6, which also indicates that the activity in stationary polymerization ceases at a mole ratio $[O_2]_0/[Ti] = 0.5$. These results have been explained in terms of the formation of $TiCl_4$ (Eq. 6.3). Doi *et al.*[11] were able to duplicate these results (Fig. 6.7 and 6.8) by using $TiCl_4$ in place of O_2 in the $(O_2 + TiCl_3) + AlEt_3$ system. The reaction of $TiCl_4$ with $AlEt_3$ results in the formation of β-type $TiCl_3$ and $AlEt_2Cl$, and the catalyst system

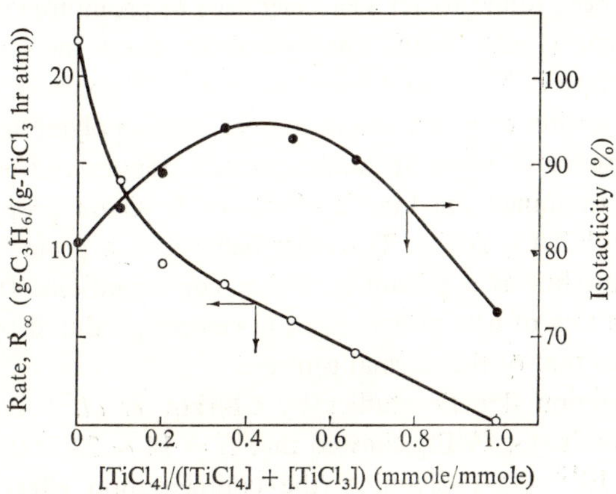

Fig. 6.7. Dependence of stationary rate and isotacticity on amount of $TiCl_4$. (40°C.)

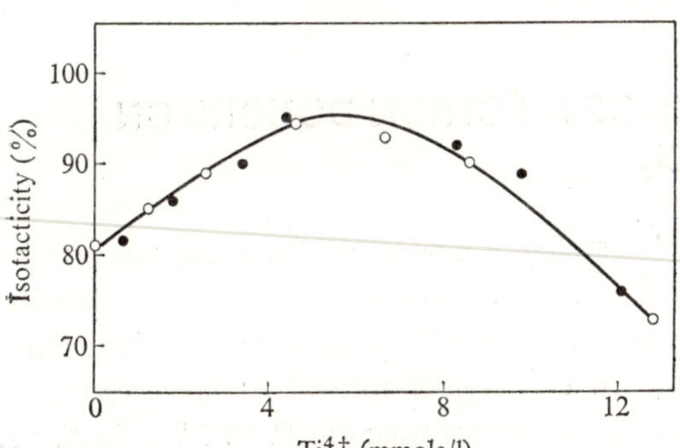

Fig. 6.8. Dependence of isotacticity on Ti^{4+} concentration. $((TiCl_3 + O_2)/AlEt_3$, ●; $(TiCl_3 + TiCl_4)/AlEt_3$, ○.)

therefore contains the following catalysts; $TiCl_4/AlEt_3$, β-$TiCl_3/AlEt_3$, α-$TiCl_3/AlEt_3$, β-$TiCl_3/AlEt_2Cl$ and α-$TiCl_3/AlEt_2Cl$. Of these, α-$TiCl_3/AlEt_3$ has the highest activity, and α-$TiCl_3/AlEt_2Cl$ has the highest stereospecificity. Doi *et al.* therefore explained their results in terms of increasing amounts of α-$TiCl_3/AlEt_2Cl$ at lower concentrations of $TiCl_4$ or O_2 (giving increasing isotacticity) and formation of more β-$TiCl_3/AlEt_2Cl$ or β-$TiCl_3/AlEt_3$ at higher $TiCl_4$ or O_2 concentrations (causing the isotacticity to decrease).

According to Soga *et al.*[10)] and Doi *et al.*,[11)] $SeOCl_2$ and O_2 can both be regarded as catalyst poisons. However, some amines and other compounds have been found to act as accelerators or promotors (effect (vi)). Thus Ambroz *et al.*[6)] found that *n*-$BuNH_2$ forms new active centres on the surface of $TiCl_3$, and Minsker *et al.*[17,18)] considered the greater ease of insertion of propylene monomer into a polymerization centre in the presence of some aromatic electron donors and ethers. Doi and Keii[19)] examined the kinetic effects of Et_2O on propylene polymerization with $TiCl_3/AlEt_3$. The ether behaves as a promoter at low concentrations, but as a poison at higher concentrations. It was found that the activity of new polymerization centres produced by the ether was 3.3 times that of the normal centres.

In this connection, kinetic studies by Chirkov *et al.*[20)] on the system ($TiCl_3$ + amine) + $AlEt_3$ showed that $E = 14 \sim 23$ kcal/mole with $A = 10^8 \sim 10^{13}$ l/(g-$TiCl_3$-min) (the compensation effect was found to hold in this range, i.e. $\ln A \propto E$) for propylene polymerization, whereas in the absence of amine, it was found that $E = 11.6$ kcal/mole and $A = 2 \times 10^6$ l/(g-$TiCl_3$ min).

6.3 INFLUENCE OF ELECTRON DONORS ON $TiCl_3/AlEtCl_2$

The $TiCl_3/AlEtCl_2$ system is a poor one for propylene polymerization, but the addition of some types of electron donor molecules containing P, S or N results in a remarkable increase both in activity and stereospecificity. The I.I. of polymer formed in the presence of the following electron donor molecules was: Bu_3P, 97.3%; Bu_2S, 97.1%; HPT, 95.5% and Bu_3N, 95.6%. These results were explained by Natta *et al.*[21)] and Zambelli *et al.*[3)] on the basis of a disproportionation reaction such as Eq. 6.2 (effect (iii)), resulting in the formation of $AlEt_2Cl$. On the other hand, McConnell *et al.*[26] concluded on the basis of I.R. measurements that disproportionation did not occur with HPT but that complex formation took place instead. Matsumura *et al.*[27] proposed the following reaction scheme to account for their findings.

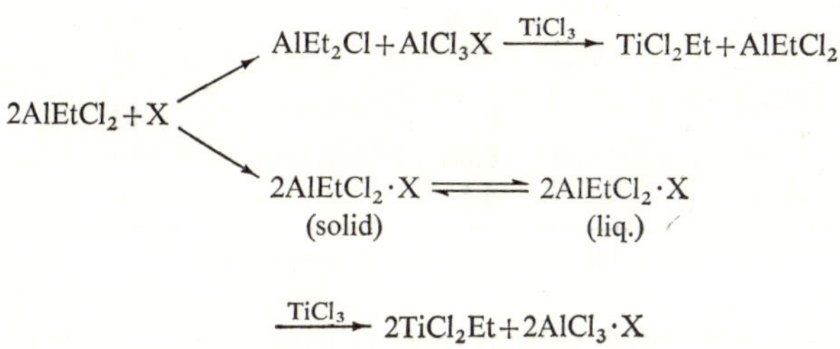

(6.8)

According to Kato *et al.*,[28] the electron donor forms a complex with $AlEtCl_2$ which is active for polymerization. This complex may be formed only at low concentrations; at higher concentration the electron donor acts as a catalyst poison.

6.4 EFFECTS OF SURFACE HETEROGENEITY

Ether behaves as a promoter for the polymerization of isoprene with $VCl_3/AlEt_3$, an effect which Cooper *et al.*[29,30] attributed to the reduction of VCl_3 to form V^{2+}, which may give rise to a polymerization centre $RVCl_2$. It has also been found by Gippin[31] that $CoCl_2/AlEt_2Cl$ acts as a catalyst for butadiene polymerization in the presence of water. The rate of polymerization and the molecular weight of the products both increase to a maximum and then decrease as the amount of water is increased.

Boor[32,33] examined the promoting effects of amines in a $TiCl_3/ZnEt_2$ catalyst. It is known that there is no interaction between the electron donors and cocatalyst in this case, so that the results, summarized in Table 6.5, could be accounted for in terms of the interactions between the amines and $TiCl_3$. A schematic representation of the dependence of conversion, isotacticity and molecular weight on the

TABLE 6.5 Influence of Et_3N on propylene polymerization with various forms of $TiCl_3$ and $ZnEt_2$†

$TiCl_3$	Et_3N/TiCl	Conversion (%)
α	0	50~70
	0.044	~30
	2.9	95
β	0	80
	0.06	48
	4.5	91
γ	0	83
	0.04	35
	4.5	70

† 50°C, 20 hr, 100 ml *n*-heptane, 27 g propylene

interaction of amines with various types of catalyst surface site is shown in Fig. 6.9. The polymerization activity (conversion) decreased to a minimum, then increased and finally decreased again on addition of further amine. At the same time, the isotacticity increased monotonically with the amount of amine. To account for this, Boor proposed the existence of several kinds of polymerization sites (Fig. 6.9).

Similar theories, postulating the presence of several types of polymerization centres, have been proposed by many authors to account for the complex effects of additives during Ziegler-Natta polymerization. Chernovskaya and Razuvaev[34] considered a heterogeneous surface of this type in discussing their data on the effects of naphthalene on propylene polymerization with α-$TiCl_3$/$AlEt_3$. Schnecko *et al.*[35] compared the activities of Ziegler-Natta catalysts poisoned by O_2 or H_2O for ethylene propylene and 1-butene polymerizations, and found that, using the same poisoned catalyst, the observed degrees of deactivation were different for each monomer. It was suggested that the polymerization centres for each monomer were therefore different.

The problem of surface heterogeneity has been discussed since the proposal of the active centre theory by Taylor[36] in 1925. Many kinds of polymerization sites have been introduced to account for particular experimental results, but the real nature of surface heterogeneity in Ziegler-Natta polymers may not be as simple as supposed by some authors. Some remarks on the heterogeneity of catalyst surfaces will be presented in the next section.

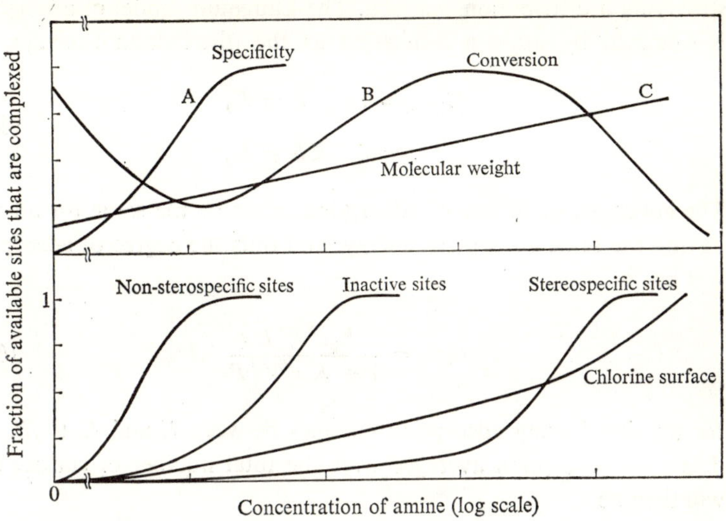

Fig. 6.9. Schematic representation of the dependence of conversion, specificity and molecular weight on the interaction of amines with different types of catalytic sites for $TiCl_3/ZnEt_2$, according to Boor.[33]

6.4.1 Surface heterogeneity and adsorption[37,38]

Let us assume a surface on which sites are distributed having various heats of adsorption, E. If the total number of surface sites is N_S and the number of sites whose adsorption energies fall between E and $E + dE$ is $N_S(E)$, we have

$$N_S = \int N_S(E) dE \tag{6.9}$$

Using the distribution function, as defined by

$$f(E) = \frac{N_S(E)}{N_S} \tag{6.10}$$

we may normalize Eq. 6.9 as follows

$$1 = \int f(E)\, dE \tag{6.11}$$

Any kind of adsorbing surface can now be characterized by

specifying a distribution function. The Langmuir model of adsorption is obtained by using a δ-function as the distribution function, i.e.

$$\begin{aligned} f(E) &= 1 \qquad E = E_0 \\ &= 0 \qquad E \neq E_0 \end{aligned} \tag{6.12}$$

The coverage, or degree of adsorption, must be the same for all sites having the same adsorption energy, and thus the degree of adsorption will be

$$\theta(E) = \frac{K_0 e^{E/RT} P}{1 + K_0 e^{E/RT} P} \tag{6.13}$$

for all sites having adsorption energies between E and $E + dE$. The degree of adsorption averaged over the total number of surface sites will then be

$$\bar{\theta} = \int \theta(E) f(E) \, dE \tag{6.14}$$

which corresponds to the observed coverage. Applying this procedure to the distribution function given in Eq. 6.12, we have a Langmuir adsorption equation

$$\bar{\theta} = \frac{K_0 e^{E_0/RT} P}{1 + K_0 e^{E_0/RT} P} \tag{6.15}$$

We may now assume a distribution function such as

$$\begin{aligned} f(E) &= f_i \qquad E = E_i \\ &= 0 \qquad E \neq E_i \\ i &= 1, 2, \ldots\ldots n \end{aligned} \tag{6.16}$$

which characterizes a surface having n types of adsorption site. Applying the normalization condition and averaging proceedure as above, we have

$$1 = \sum_{i=1}^{n} f_i \tag{6.17}$$

$$\bar{\theta} = \sum_{i=1}^{n} f_i\left(\frac{K_0 e^{E_i/RT}P}{1 + K_0 e^{E_i/RT}P}\right)$$

We may suppose, as a special case, that

$$E_1 \gg E_2 = E_3 = \ldots\ldots = E_n \equiv E \tag{6.16a}$$

The corresponding adsorption equation is then

$$\bar{\theta} = f_1 + (1 - f_1)\frac{K_0 e^{E/RT}P}{1 + K_0 e^{E/RT}P} \tag{6.17a}$$

The two distribution functions given below have been considered as special cases of continuous distribution functions.

$$f(E) = \text{const.}\left(\equiv \frac{1}{u}\right) \tag{6.18}$$

$$f(E) = \beta e^{-\beta E} \tag{6.19}$$

Eq. 6.18 gives rise to a Frumkin-Temkin equation

$$\bar{\theta} = \left(\frac{RT}{u}\right) \ln (K_0 e^{E_0/RT}P) \tag{6.20}$$

where E_0 is the upper limit of the adsorption energy, and $U\ (= E_0 - E_t)$ is the range of adsorption energies. Eq. 6.19 gives approximately a Freundlich equation.

$$\bar{\theta} = (K_0 P)^{\beta RT} \tag{6.21}$$

The heat of adsorption $\bar{E}$ on a heterogeneous surface is given by

$$\bar{E} = \frac{1}{\bar{\theta}}\int E\,\theta(E) f(E)\,dE \tag{6.22}$$

from application of the Clausius-Clapeyron equation for theoretical models such as Eq. 6.20 or 6.21. Fig. 6.10 shows the relationship between heat of adsorption and coverage for heterogeneous surfaces based on the above models.

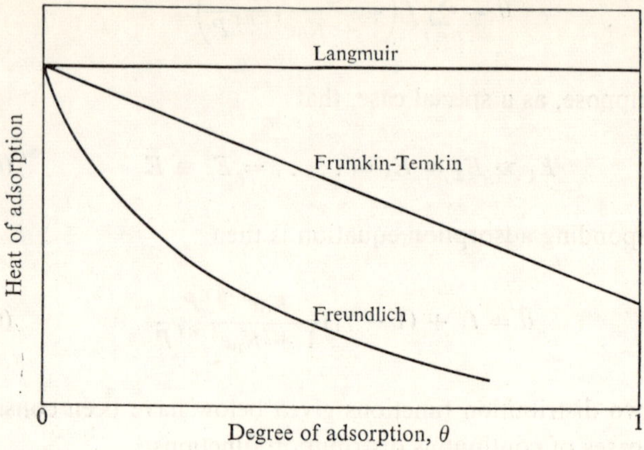

Fig. 6.10. Theoretical relationship between heat of adsorption and degree of adsorption for various heterogeneous surfaces.

6.4.2 Catalyst surface heterogeneity

Let us assume a catalyst surface on which active centres (or polymerization centres) are distributed, with activation energy E^*. We may define the distribution function as before.

$$f(E^*) = \frac{N_S(E^*)}{N_S} \tag{6.23}$$

$$1 = f(E^*)dE^* \tag{6.24}$$

The observed rate of reaction can be regarded as the averaged rate over all the centres. Provided that the rate of reaction over active sites of energy E^* can be expressed as

$$\begin{aligned} R &= k_0' e^{-E^*/RT}[\text{M}][N_S(E^*)] \\ &= N_S k_0 e^{-E^*/RT} f(E^*) \end{aligned} \tag{6.25}$$

the averaged rate will be

$$\bar{R} = N_S \int k_0 e^{-E^*/RT} f(E^*) dE^* \tag{6.26}$$

Applying the distribution function for a homogeneous surface (i.e. Langmuir type)

$$\begin{aligned} f(E^*) &= 1 \qquad E^* = E_0^* \\ &= 0 \qquad E^* \neq E_0^* \end{aligned} \tag{6.27}$$

we obtain

$$\bar{R} = N_S k_0' e^{-E_0^*/RT}[\mathrm{M}] \tag{6.28}$$

For a non-uniform surface with the distribution function

$$\begin{aligned} f(E^*) &= f_i^* \qquad E^* = E_i^* \\ &= 0 \qquad E^* \neq E_i^* \\ i &= 1, 2, \ldots, n \end{aligned} \tag{6.29}$$

the averaged rate becomes

$$\begin{aligned} \bar{R} &= N_S[\mathrm{M}] \sum_{i=1}^{n} f_i^* k_0' e^{-E_i^*/RT} \\ 1 &= \sum_{i=1}^{n} f_i^* \end{aligned} \tag{6.30}$$

As a special case of Eq. 6.29, we may suppose that

$$E_1^* \ll E_2^*, E_3^*, \ldots. E_n^* \tag{6.29a}$$

which gives approximately the following rate.

$$\bar{R} \simeq N_S f_1^* k_0' e^{-E_1^*/RT}[\mathrm{M}] \tag{6.30a}$$

The last result (Eq. 6.30*a*) can be expected to hold for a catalyst surface having a small number of active centres, and provides an explanation for the small frequency factors observed in some heterogeneous reactions ($1 \gg f_i^*$), and for catalyst deactivation by small amounts of poisons.

If we assume a distribution function of the form

$$f(E^*) = \beta e^{\beta E^*} \tag{6.31}$$

the following rate law can be derived.

$$\bar{R} \simeq N_S f(E^*_{\min}) k'_0 e^{-E^*_{\min}/RT}[\mathrm{M}] \tag{6.32}$$

where $E^*_{\min}$ is the lower limit of the activation energy. In general a catalytic poison may be expected to deactivate the most active centres with the lowest activation energies initially, with subsequent deactivation of the less active centres of higher energy. It may therefore be anticipated that the apparent value of $E^*_{\min}$ will increase with increasing amount of poison. This accounts for the "compensation effect" found by Chirkov *et al.*,[20] i.e.

$$\ln A \propto \ln f(E^*_{\min}) \propto \beta E^*_{\min} \propto E^*_{\mathrm{obs}} \tag{6.33}$$

6.4.3 Surface heterogeneity and molecular weight distribution

It is generally accepted that the molecular weight distribution of polyolefins produced with Ziegler-Natta (or Phillips) catalysts is very broad.[39−41] The reasons for this wide distribution have not yet been clearly established, though two main approaches to the problem have been made.[42] One treatment has been worked out by Clark and Bailey,[43] who considered ethylene polymerization on a non-uniform surface. The second treatment is due to Gordon and Roe,[44,45] who assumed that the rate constants of chain termination varied with chain length. Grieveson,[46] and Berger *et al.*[42] investigated the molecular weight distribution of polyethylene produced with γ-$TiCl_3$/$AlEt_2Cl$/H_2 and supported the former approach. The theory of Clark and Bailey will be outlined below.

Number and weight average molecular weights ($\overline{M}_n$ and $\overline{M}_w$ respectively) may be defined as

$$\overline{M}_n = M\frac{\Sigma\, nN_n}{\Sigma\, N_n} \tag{6.34}$$

$$\overline{M}_w = M\frac{\Sigma\, n^2N_n}{\Sigma\, nN_n} \tag{6.35}$$

where N_n is the number of molecules containing n monomer units and M is the molecular weight of the monomer. The summations extend from $n = 2$ to $n = \infty$.

A. The Rideal model

When the rate of polymerization reaches the stationary state, the number of polymer molecules desorbed per unit catalyst surface per second is given by

$$\sum N_n = k_d N_0 \sum_{n=2}^{\infty} \theta_n \tag{6.36}$$

where k_d, the desorption rate constant, is the number of polymer molecules desorbed per site per second, N_0 is the number of sites per unit surface area, and θ_n is the fraction of sites covered with n-mer. It is assumed that k_d is independent of n. Similarly, we have the following equations

$$\sum nN_n = k_d N_0 \sum_{n=2}^{\infty} n\theta_n \tag{6.37}$$

$$\sum n^2 N_n = k_d N_0 \sum_{n=2}^{\infty} n^2\theta_n \tag{6.38}$$

Under steady state conditions, consideration of adsorbed n-mer gives

$$k_r P\theta_{n-1} = (k_r P + k_d)\theta_n \tag{6.39}$$

where k_r (assumed independent of n) is the rate constant for the reaction of gas phase monomer with adsorbed n-mer, and P is pressure of monomer. From Eq. 6.39, we have

$$\begin{aligned}\sum_{n=2}^{\infty} \theta_n &= \sum_{n=1}^{\infty} \theta_n - \theta_1 \\ &= \left(\frac{k_r P}{k_d}\right)\theta_1\end{aligned} \tag{6.40}$$

$$\sum_{n=2}^{\infty} n\theta_n = \frac{k_r P(k_r P + 2k_d)}{k_d^2}\theta_1 \tag{6.41}$$

$$\sum_{n=2}^{\infty} n^2\theta_n = \frac{k_r P(2k_r^2 P^2 + 5k_r P k_d + 4k_d^2)}{k_d^3}\theta_1 \tag{6.42}$$

The term θ_1 is evaluated explicitly from the equilibrium equation for adsorbed monomer

$$k_a P\left(1 - \sum_{n=1}^{\infty} \theta_n\right) = (k_r P + k_d)\theta_1 \tag{6.43}$$

where k_a is the monomer adsorption rate constant and the term in parentheses on the left hand side represents the fraction of unoccupied sites. Expressions for the number average and weight average molecular weights, their ratio, and the rate of polymerization can now be derived, putting $r \equiv k_r P/k_d$.

$$\overline{M}_n/M = r + 2 \tag{6.44}$$

$$\overline{M}_w/M = \frac{2r^2 + 5r + 4}{r + 2} \tag{6.45}$$

$$\overline{M}_w/\overline{M}_n = \frac{2r^2 + 5r + 4}{(r + 2)^2} \tag{6.46}$$

$$\begin{aligned} \bar{R} = \sum_{n=2}^{\infty} nN_n &= \frac{N_0 k_a P^2 k_r (k_r P + 2k_d)}{(k_a P + k_d)(k_r P + k_d)} \\ &= \frac{N_0 k_a P r(r + 2)}{(1 + (k_a/k_d)P)(r + 1)} \end{aligned} \tag{6.47}$$

From these results, it was concluded that a rather narrow molecular weight distribution

$$1 \leqslant \overline{M}_w/\overline{M}_n \leqslant 2 \tag{6.48}$$

could be anticipated for a homogeneous surface.

Clark and Bailey[43] further developed this approach, and considered the case of a non-uniform (heterogeneous) surface having the following normalized distribution function.

$$f(E) = ce^{-cE} \tag{6.49}$$

Eq. 6.40 to 6.42 can be rewritten under this condition as follows

$$\sum N_n = \int_0^{\infty} N_0 k_d(E) \sum_{n=2}^{\infty} \theta_n(E) f(E) dE \tag{6.50}$$

$$\sum nN_n = \int_0^{\infty} N_0 k_d(E) \sum_{n=2}^{\infty} n\theta_n(E) f(E) dE \tag{6.51}$$

$$\sum n^2 N_n = \int_0^{\infty} N_0 k_d(E) \sum_{n=2}^{\infty} n^2\theta_n(E) f(E) dE \tag{6.52}$$

and the resulting expressions obtained by the authors for the molecular weights, their ratio, and the polymerization rate were

$$\frac{\overline{M}_n}{M} = \frac{\left[\dfrac{s-q}{s}\displaystyle\int_0^1 \frac{Z^{p-1}}{1+sZ}dZ + \int_0^1 \frac{Z^{p-1}}{1+qZ}dZ\right]}{\left[\displaystyle\int_0^1 \frac{Z^{p-1}}{1+qZ}dZ - \int_0^1 \frac{Z^{p-1}}{1+sZ}dZ\right]} \tag{6.53}$$

$$\frac{\overline{M}_w}{M} = \frac{\left[\dfrac{2(s-q)}{q^2}\displaystyle\int_0^1 Z^{p-2}dZ + \frac{5sq-4q^2-2s^2}{q^2}\int_0^1 \frac{Z^{p-1}}{1+sZ}dZ + \int_0^1 \frac{Z^{p-1}}{1+qZ}dZ\right]}{\left[\dfrac{s-q}{s}\displaystyle\int_0^1 \frac{Z^{p-1}}{1+sZ}dZ + \int_0^1 \frac{Z^{p-1}}{1+qZ}dZ\right]} \tag{6.54}$$

$$\frac{\overline{M}_w}{\overline{M}_n} = \frac{\left[\dfrac{2(s-q)}{q^2}\displaystyle\int_0^1 Z^{p-2}dZ + \frac{5sq-4q^2-2s^2}{q^2}\int_0^1 \frac{Z^{p-1}}{1+sZ}dZ + \int_0^1 \frac{Z^{p-1}}{1+qZ}dZ\right]}{\left[\left(\dfrac{s-q}{s}\right)\displaystyle\int_0^1 \frac{Z^{p-1}}{1+sZ}dZ + \int_0^1 \frac{Z^{p-1}}{1+qZ}dZ\right]^2}$$
$$\times \left[\int_0^1 \frac{Z^{p-1}}{1+qZ}dZ - \int_0^1 \frac{Z^{p-1}}{1+sZ}dZ\right] \tag{6.55}$$

$$\overline{R} = const. \left[\left(\frac{s-q}{s}\right)\int_0^1 \frac{Z^{p-1}}{1+sZ}dZ + \int_0^1 \frac{Z^{p-1}}{1+qZ}dZ\right] \tag{6.56}$$

The following relationships hold for the above equations.

$$k_d(E) = \beta e^{-E/RT} \qquad Z = e^{-E/RT}$$

$$s = \frac{B}{k_a P} \qquad const\ (\text{from } \overline{R}) = \frac{k_r spPN_0}{s-q} \tag{6.57}$$

$$q = \frac{B}{k_r P} \qquad p = RTC$$

These results indicate that the ratio $\overline{M}_w/\overline{M}_n$ changes with the value of k_a/k_r, as shown in Fig. 6.11. The broad molecular weight distribution that is observed experimentally can thus be explained in terms of catalyst surface heterogeneity. Clark and Bailey also considered the Langmuir-Hinshelwood model, and reached similar conclusions.

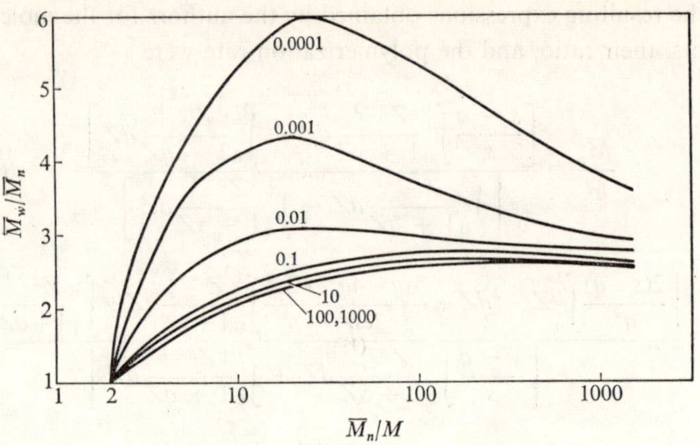

Fig. 6.11. Weight average/number average molecular weight ratio *vs.* $(1/M) \times$ number average molecular weight with k_a/k_r as a parameter. Rideal mechanism, exponential distribution. $p = RTC = 2$.

References

1. D.F. Hoeg, S. Liebman and L. Schubert, *J. Org. Chem.*, **28**, 1554 (1963).
2. H.E. Swift, C.P. Poole Jr. and J.F. Itzel Jr., *J. Phys. Chem.*, **68**, 2509 (1964).
3. A. Zambelli, J.D. Pietro and G. Gatti, *J. Polymer Sci.*, **A1**, 403 (1963).
4. K. Tanabe and Y. Watanabe, *Shokubai* (Japanese), **4**, 360 (1962).
5. S. Kanetaka, *Kogyo Kagaku Zasshi* (Japanese), **70**, 1568 (1967).
6. J. Ambroz and O. Hamerik. *Collection Czech. Chem. Commun.*, **28**, 2550 (1963).
7. H. Martin and E. Blanck. *Makromol. Chem.*, **69**, 1 (1963).
8. K. Vessely, J. Ambroz, R. Vilim and O. Hamerik, *J. Polymer Sci.*, **55**, 25 (1961).
9. E.B. Milovskaya and P.I. Polgopolskaya, *Vysokomolekul. Soedin.*, **4**, 147 (1962).
10. K. Soga, Y. Takano, S. Go and T. Keii *J. Polymer Sci.*, **A5**, 2815 (1967).
11. Y. Doi, Y. Hattori, I. Okura and T. Keii, *Kogyo Kagaku Zasshi* (Japanese), **72**, 2621 (1969).
12. A.K. Ingberman, *J. Polymer Sci.*, **A4**, 2781 (1966).
13. K. Ziegler, F. Krupp and K. Zosel, *Ann. Chem.*, **629**, 241 (1960).
14. M.H. Abraham and A.G. Davies, *J. Chem. Soc.*, **1959**, 429.
15. Y. Nakayama, T. Tsuruta and J. Furukawa, *Makromol. Chem.*, **40**, 79 (1960).
16. C. Beermann and H. Bestian, *Angew. Chem.*, **71**, 618 (1959).
17. R.P. Chernovskaya, K.S. Minsker and G.A. Ruzukaev, *Vysokomolekul. Soedin.*, **6**, 1656 (1964).
18. G.A. Razuvaev, K.S. Minsker, G.T. Fedoseeva and V.K. Bykhovskii, *Polymer Sci. USSR*, **2**, 299 (1961).
19. Y. Doi and T. Keii, *Kogyo Kagaku Zasshi* (Japanese), **73(11)**, 2468 (1970).
20. O.N. Pirogov and N.M. Chirkov, *Polymer Sci.*, **7(3)**, 544 (1965); *Vysokomolekul. Soedin.*, **7**, 491 (1965).
21. G. Natta, A. Zambelli, I. Pasquon, G. Gatti and D. De Luca, *Makromol. Chem.*, **70**, 206 (1964).

22. A. Nicco, *French Pat.* 1354815 (1964).
23. H.W. Coover Jr. and F.B. Joyner, *J. Polymer Sci.*, **A3**, 2407 (1965).
24. Toyo Rayon Co., *Brit. Pat.* 1017708 (1966).
25. W.R. Wattaand and C.D. Fisher, *J. Polymer Sci.*, **A7**, 2815 (1969).
26. R.L. McConnell, M.A. McCall, G.O. Cash Jr., F.B. Joyner and H.W. Coover Jr., *J. Polymer Sci.*, **A3**, 2135 (1965).
27. K. Matsumura, Y. Atarashi and O. Fukumoto, *J. Polymer Sci.*, **A7**, 311 (1969).
28. A. Kato, Y. Doi, I. Okura and T. Keii, presented at Annual Meeting of Japan Chem. Soc. (1971).
29. W. Cooper, D.E. Eaves, G.D.T. Owen and G. Vaughan, *J. Polymer Sci.*, **C4**, 211 (1963).
30. W. Cooper, R.K. Smith and A. Stokes, *J. Polymer Sci.*, **B4**, 309 (1966).
31. N. Gippin, *Ind. Eng. Chem. Prod. Res. Develop.*, **1**, 32 (1962).
32. J. Boor Jr., *J. Polymer Sci.*, **C1**, 257 (1963).
33. J. Boor Jr., *J. Polymer Sci.*, **A3**, 995 (1965).
34. R.P. Chernovskaya and G.A. Razuvaev, *Vysokomolekul. Soedin.*, **6**, 1656 (1964).
35. H. Schnecko, W. Linty and W. Kern, *J. Polymer Sci.*, **A5**, 205 (1967).
36. H.S. Taylor, *Proc. Roy. Soc.*, **A108**, 105 (1925).
37. S.Z. Roginskii, *Adsorption and Catalysis on Non-uniform Surfaces* (Russian), 1948.
38. T. Keii, *Shokubai* (Japanese), **13**, 29 (1956); *J. Chem. Phys.*, **25**, 1283 (1956); *J. Res. Inst. Catalysis, Hokkaido Univ.*, **7**, 99 (1959).
39. H. Wesslau, *Makromol. Chem.*, **20**, 111 (1956); 26, 102 (1958).
40. L.H. Tung, *J. Polymer Sci.*, **24**, 333 (1957).
41. G.K. Williamson, B. Wright and R.N. Haward, *J. Appl. Chem.*, **14**, 131 (1964).
42. M.N. Berger, G. Boocock and R.N. Haward, *Advan. Catalysis*, **19**, 211 (1968).
43. A. Clark and G.C. Bailey, *J. Catalysis*, **2**, 230 (1963).
44. M. Gordon and R.J. Roe, *Polymer*, **2**, 41 (1961).
45. R.J. Roe, Polymer, **2**, 60 (1961).
46. B.M. Grieveson, *Makromol. Chem.*, **84**, 93 (1965).

Chapter 7

Some related reactions of Ziegler-Natta catalysts

In this chapter, some further reactions of Ziegler-Natta catalysts will be considered as a basis for the discussion in chapter 9 of the reaction mechanism in Ziegler-Natta polymerization.

7.1 GAS-PHASE POLYMERIZATION WITH $TiCl_3$/ALKYLALUMINIUM

7.1.1 Gas-phase polymerization with $TiCl_3$

Keii *et al.*[1)] found that $TiCl_3$ catalysts (H-, HA-, A- and AA-grades) show some activity in the gas-phase polymerization of propylene in the absence of any cocatalyst, even at low temperatures and pressures. The activities were in the following order

$$AA \gg HA > A > H \tag{7.1}$$

which is in agreement with the order of their acidity, reported by Tanabe *et al.*[2)] The molecular weight of the product was 500–1500 when the reaction was carried out at room temperature and low pressure (initially 20 cm Hg).

It was found that the activity of H-$TiCl_3$ could be increased by the addition of small amounts of $AlCl_3$ (< 10%). This catalyst then gave the same ESR spectrum as AA-$TiCl_3$[3)], so that the high activity of AA-$TiCl_3$ may be regarded as due to the presence of $AlCl_3$ in the surface layer. Soga *et al.*[4)] found that the high activity of AA-$TiCl_3$ could be depressed by washing the sample with toluene, and at the same time it was confirmed that a considerable amount of $AlCl_3$ was dissolved into

TABLE 7.1 Cationic polymerization by various samples of $TiCl_3$

Samples of $TiCl_3$	Washing	$TiCl_3$ used (g)	Surface area (m^2)	Stationary rate (ml_{STP}/(hr m^2))
Stauffer HA	—	1.00	14.1	0.085
Stauffer AA	—	1.28	32.0	0.224
Stauffer AA	toluene	0.62	40.3	0.089
Stauffer AA	toluene	0.71	46.1	0.083
Stauffer AA	*n*-heptane†	0.95	61.7	0.163
Stauffer AA	*n*-heptane	1.10	71.4	0.143

† Transferred quantity of $AlCl_3$: $0 \sim 4.6 \times 10^{-6}$ mole/(100 g *n*-heptane g-$TiCl_3$).

the toluene (30×10^{-6} mole $AlCl_3$/(100 g-toluene g-$TiCl_3$)). The change in polymerization activity per unit surface area on washing the catalysts with good or bad solvents of $AlCl_3$ is shown in Table 7.1, which confirms that the activity for gas-phase polymerization is due to $AlCl_3$ in the catalyst surface layer.

7.1.2 Gas-phase polymerization with $TiCl_3/AlEt_3$ and $TiCl_3/AlEt_2Cl$

The gas-phase polymerization of propylene using $TiCl_3/AlEt_3$ and $TiCl_3/AlEt_2$ has been investigated by Doi *et al.*[5] using the reactor illustrated in Fig. 7.1. In order to prevent the coalescence of dry catalyst particles the reactor contained about 10 ml of small glass beads (diameter 1.5 mm) which were agitated by a magnetic stirrer. A given amount of $TiCl_3$ and cocatalyst (10 wt% *n*-heptane solution) was admitted into the reactor, which was then evacuated for one hour in order to remove the solvent completely. Propylene was then introduced, with a stirring speed of 400 rpm, and the kinetic curves shown in Fig. 7.2 were obtained at various mole ratios of catalyst to cocatalyst. The upper curves were obtained in the presence of 150 ml *n*-heptane (normal liquid-phase polymerization). It was confirmed that the polymerization rate was proportional to the mass of $TiCl_3$ at stirring speeds above 200 rpm.

As can be seen from Fig. 7.2, the general form of the kinetic curves obtained during gas-phase polymerization with $TiCl_3/AlEt_3$ is the same as that found during liquid-phase polymerization, though the maximum is less pronounced. In the case of $TiCl_3/AlEt_2Cl$, the gas- and liquid-

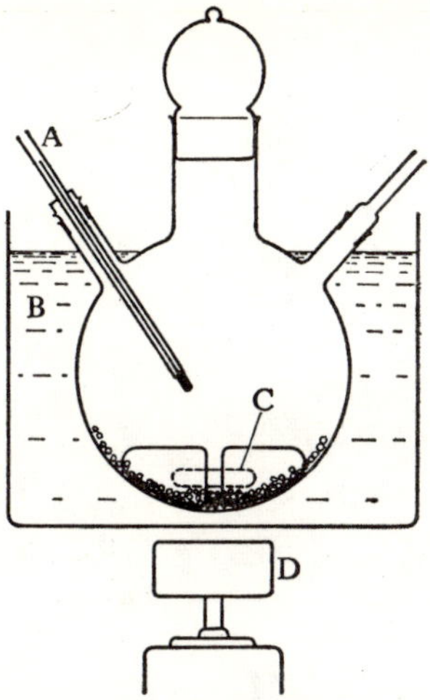

Fig. 7.1. Reactor used for gas-phase polymerization. A, thermometer; B, thermostat; C, Teflon covered magnetic stirrer; D, magnet.

phase kinetic curves are again similar, but here the stationary rate, as well as the rate maximum, is lower for gas-phase polymerization. The dependence of the stationary rate on the mole ratio is similar for both catalyst systems and is illustrated in Fig. 7.3 for $TiCl_3/AlEt_2Cl$. The saturation values at high mole ratios are equal to the corresponding values for liquid-phase polymerization, though the latter are reached at lower mole ratios ([A]/[Ti] $\simeq$ 1). The dependence of stationary rate on propylene pressure at various mole ratios is shown in Fig. 7.4. The curves are calculated from Eq. 7.2.

$$R_\infty = \frac{kK(A)P^2}{1 + K(A)P^2}$$

$$= R_\infty^L\left(\frac{K(A)P}{1 + K(A)P}\right) \tag{7.2}$$

where (A) is the mole ratio, P is the propylene pressure and R_∞^L is the corresponding saturation value of the rate for liquid-phase polymerization. The values of the constants k and K are 20.4 and 0.5, respectively, at 53 °C. Fig. 7.5 shows a plot of k and K *vs.* temperature, from which we obtain $k = k_0 e^{-11\text{kcal}/RT}$ and $K = K_0 e^{9.3\text{kcal}/RT}$.

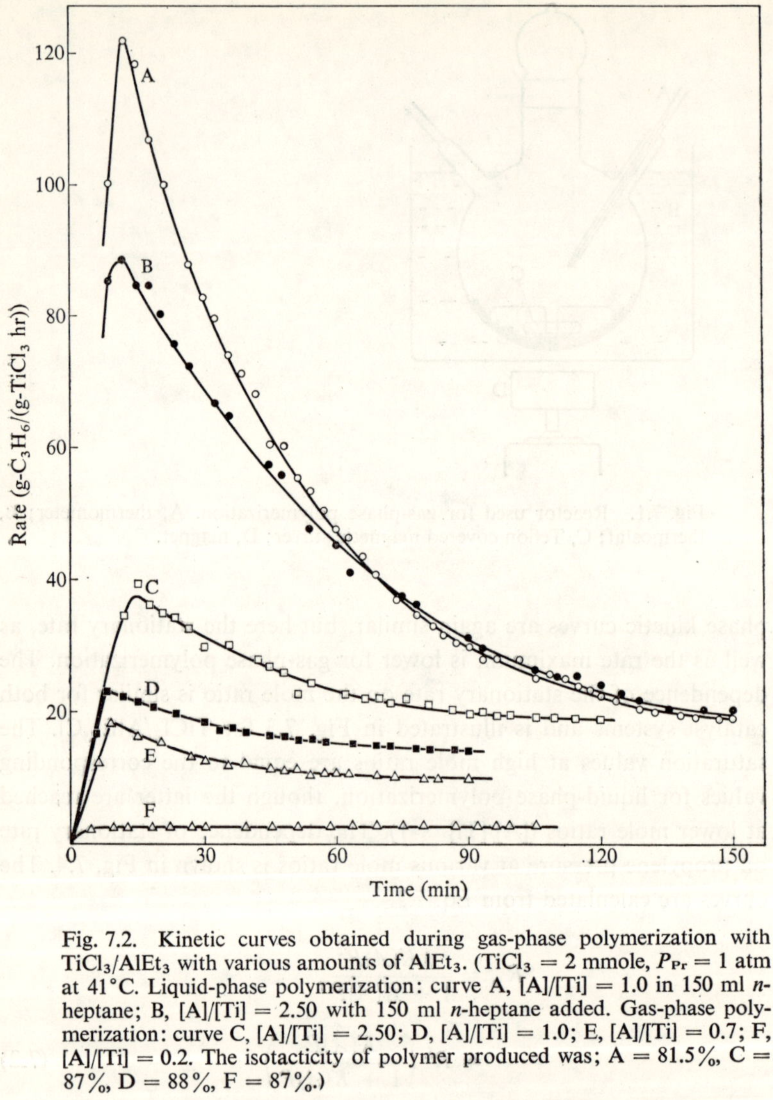

Fig. 7.2. Kinetic curves obtained during gas-phase polymerization with $TiCl_3/AlEt_3$ with various amounts of $AlEt_3$. ($TiCl_3$ = 2 mmole, P_{Pr} = 1 atm at 41°C. Liquid-phase polymerization: curve A, [A]/[Ti] = 1.0 in 150 ml *n*-heptane; B, [A]/[Ti] = 2.50 with 150 ml *n*-heptane added. Gas-phase polymerization: curve C, [A]/[Ti] = 2.50; D, [A]/[Ti] = 1.0; E, [A]/[Ti] = 0.7; F, [A]/[Ti] = 0.2. The isotacticity of polymer produced was; A = 81.5%, C = 87%, D = 88%, F = 87%.)

Eq. 7.2 is applicable to both catalyst systems, but it should be noted that the isotacticity of the polymer is not the same. For $TiCl_3/AlEt_2Cl$, the isotacticity is the same as that observed for the corresponding liquid-phase polymerization, whereas for $TiCl_3/AlEt_3$ the isotacticity is higher in the gas-phase, as shown in Fig. 7.6. This enhancement of isotacticity

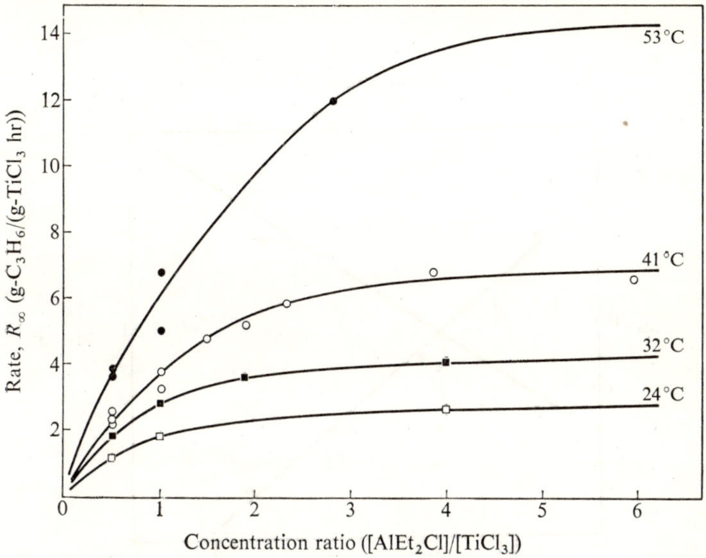

Fig. 7.3. Dependence of stationary rate on mole ratio at various temperatures. Gas-phase polymerization with $TiCl_3/AlEt_2Cl$.

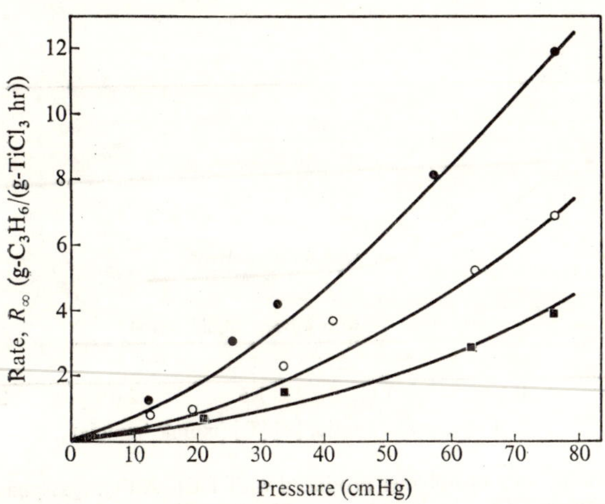

Fig. 7.4. Dependence of stationary rate on propylene pressure at various mole ratios for gas-phase polymerization with $TiCl_3/AlEt_2Cl$ at 53°C. (Curves calculated from Eq. 7.2, with $k = 20.4$ and $K = 0.5$. [A]/[Ti] = 2.74, ●; [A]/[Ti] = 1.0, ○; [A]/[Ti] = 0.5, ■.)

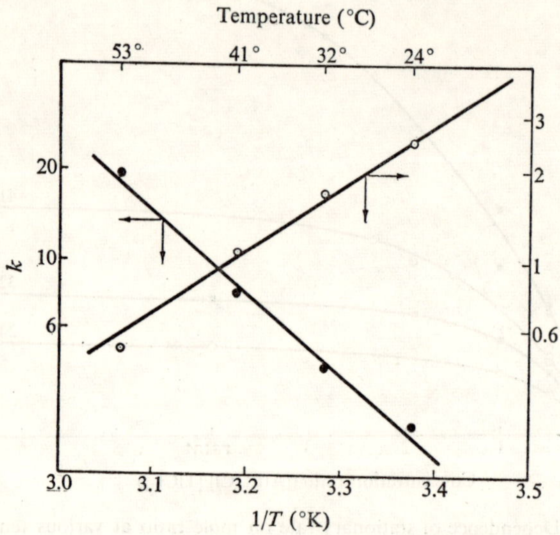

Fig. 7.5. Variation of k and K (Eq. 7.2.) with temperature.

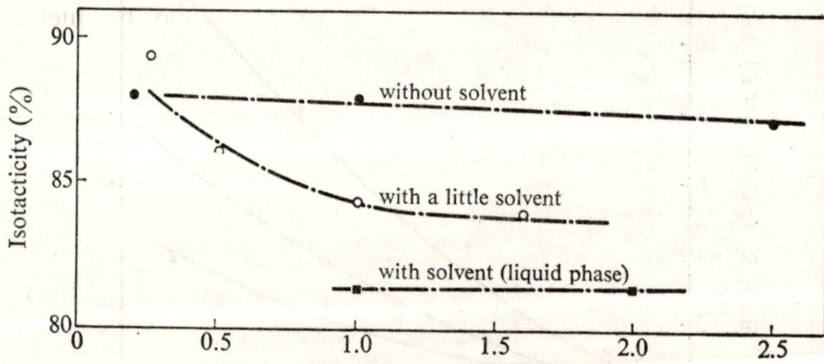

Fig. 7.6. Isotacticity of polymer produced with $TiCl_3/AlEt_3$. (gas-phase, ●; 1 ml n-heptane/g-$TiCl_3$, ○; liquid-phase, ■.)

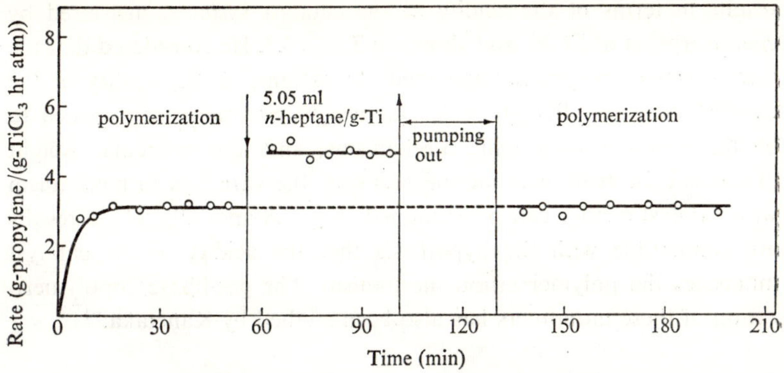

Fig. 7.7. Effect of solvent on gas-phase polymerization. ($TiCl_3/AlEt_2Cl$, with [A]/[Ti] = 1.0.)

may be understood in terms of the reciprocal relationship between rate and integral isotacticity, as pointed out in 4.2.4.

The experiment results summarized by Eq. 7.2 are rather difficult to explain. Curve B of Fig. 7.2 shows that the effect of solvent addition after initial evacuation of the reaction system is to increase the polymerization rate (*cf.* curve C), and this effect is reversible, as can be seen from Fig. 7.7. It may therefore be supposed that the excess cocatalyst at higher mole ratios acts in some way as a solvent. The depression of the rate observed in gas-phase polymerization may be due to the blocking of active centres by polymer, and the rate recovery with increasing amount of liquid cocatalyst, or on addition of solvent, may be due to the removal of this polymer leading to an increase in the rate of formation of polymerization centres from the active centres. The formation of polymerization centres requires the participation of propylene monomer and cocatalyst, as described in 2.2, resulting in a mixed first/second order dependence of the rate on propylene pressure, and on this basis Eq. 7.2 can be readily derived.

7.1.3 Gas-phase α-olefin polymerization

Kanetaka[6] examined and compared the gas-phase polymerization of ethylene, propylene and isobutene by $TiCl_3$ and alkylaluminium ($AlEt_3$, $AlEt_2Cl$, $AlEtCl_2$) in the presence and absence of some amines. The results are summarized in Table 7.2. Kanetaka discussed these results in terms of the acidity of the catalyst systems, measured by chemisorption of Et_3N, and shown in Table 7.3. He considered that the mechanism of polymerization tends to cationic as the acidity of the catalyst increases. If high molecular weight polymers are assumed to be the products of anionic mechanisms, and low molecular weight products arise from cationic mechanisms, the variation in mechanism with catalyst acidity can be examined (Fig. 7.8 and 7.9). These results are compatible with the hypothesis that the acidity of the catalyst influences the polymerization mechanism. The gas-phase copolymerization of these monomers has also been studied by Kanetaka.[6)]

TABLE 7.2 Gas phase polymerization of α-olefin[†1]

Cocatalyst (Al/Ti)		Olefin	Polymer produced at 2 hr	Amount of product at 2 hr (g/g-$TiCl_3$)
$AlEt_3$	(0.59)	C_2H_4	high polymer[†2]	8
$AlEt_2Cl$	(0.57)	C_2H_4	high polymer	1.7
$AlEt_2Cl$	(0.29)	C_2H_4	high polymer	0.6
$AlEtCl_2$	(0.46)	C_2H_4	low polymer[†3]	0.2
$AlEtCl_2$	(0.81)	C_2H_4	low polymer	0.31
$AlEt_3$	(0.65)	C_3H_6	high polymer	6.9
$AlEt_2Cl$	(0.46)	C_3H_6	high polymer	1.0
$AlEtCl_2$	(0.36)	C_3H_6	low polymer	1.0
$AlEt_3$	(0.71)	i-C_4H_8	—	0
$AlEt_2Cl$	(0.51)	i-C_4H_8	low polymer	0.2
$AlEt_2Cl$	(0.75)	i-C_4H_8	low polymer	0.6
$AlEtCl_2$	(0.54)	i-C_4H_8	low polymer	14.0
$AlEtCl_2$ + Et_3N	(0.5) (0.3)[†4]	C_2H_4	high polymer	0.4
$AlEtCl_2$ + Et_3N	(0.5) (0.5)[†4]	C_3H_6	high polymer	0.06
$AlEtCl_2$ + Et_3N	(0.35) (0.08)[†4]	C_3H_6	low polymer[†5]	0.85
$AlEtCl_2$ + Et_3N	(0.50) (0.25)[†4]	iso-C_4H_8	low polymer	8.2
$AlEtCl_2$ + Et_3N	(0.50) (0.3)[†4]	C_3H_6	—	0

†1 40°C, 40 cmHg, $TiCl_3$ (AA, Stauffer Chem. Co., 40 m^2/g).
†2 High polymer, $\overline{M}_w = 3{\sim}4 \times 10^5$.
†3 Low polymer, $\overline{M}_w = 200{\sim}500$.
†4 Figures in parentheses indicate $Et_3N/AlEtCl_2$.
†5 Rubberlike.

TABLE 7.3 Acidity of $TiCl_3$-alkylaluminium systems

Catalyst system (Al/Ti)	Adsorption of Et_3N (20°C)		Phenol adsorption (20°C)	
	Physical adsorption[†1] (mmole/g)	Chemisorption[†2] (mmole/g) ≡ acidity	Physical adsorption (mmole/g)	Chemisorption (mmole/g) ≡ basicity
$TiCl_3$ (AA)	0.17	2.99		
$TiCl_3$–$AlEtCl_2$ (0.75)	0	3.40		
$TiCl_3$–$AlEtCl_2$ (1.055)	0	4.60		
$TiCl_3$–$AlEt_2Cl$ (0.385)	0.39	2.00		
$TiCl_3$–$AlEt_2Cl$ (1.04)	2.62	0.16		
$TiCl_3$–$AlEt_3$ (0.63)	0.79	0.545		
$TiCl_3$–$AlEt_3$ (1.05)			0	0.445

†1 Amount desorbed by evacuation for 1 hr after adsorption equilibrium was established. Desorption was very rapid and finished within 1 min at this temperature.

†2 Adsorbed material remaining after 1 hr evacuation.

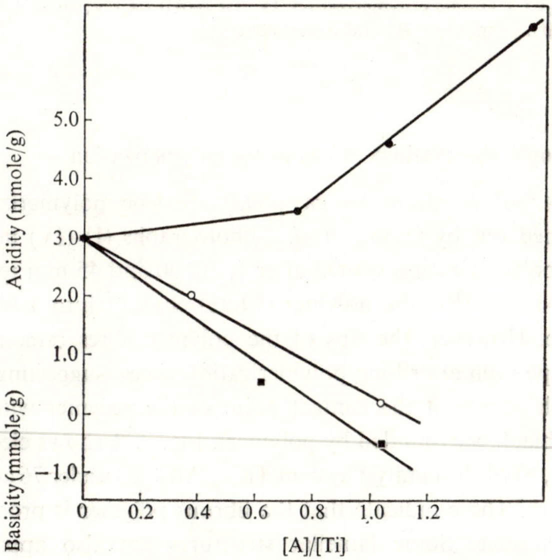

Fig. 7.8. Acidity of some catalyst systems. ($TiCl_3/AlEt_3$ (■); $TiCl_3/AlEt_2Cl$ (○); $TiCl_3/AlEtCl_2$ (●).)

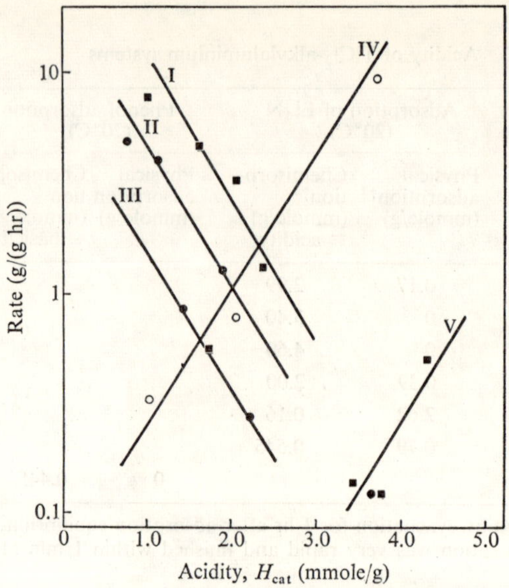

Fig. 7.9. Relationship between catalyst acidity and activity in anionic or cationic polymerization (Anionic polymerization: I, II, $TiCl_3/AlEt_3$; III, $TiCl_3/AlEt_2Cl$. Cationic polymerization: IV, $TiCl_3/AlEt_2Cl$; V, $TiCl_3/AlEtCl_2$. Ethylene (●), propylene (■) and isobutylene (○).)

7.1.4 Microscopic observations of gas-phase polymerization

Microscopic observations of gas-phase ethylene polymerization have been carried out by Okura *et al.*[7] Photographs (I)–(IV) in Fig. 7.10 show the polymerization system after 5, 20, 30 and 45 min respectively. It can be seen that the polymer (white) changes into a fibrous state with time. However, the tips of the polymer fibres remain unchanged in shape even after long polymerization times, suggesting that polymer growth occurs at the contact point of the polymer with the $TiCl_3$ surface (black, surrounded by polymer). Fig. 7.11 shows electron micrographs (a, b) of the catalyst system $TiCl_3/AlEt_2Cl$ under 76 cm Hg ethylene pressure. These indicate that the fibrous polymer is produced at the $TiCl_3$ surface. Some lamellar structures are also apparent. Photograph b shows polymer treated for 24 hr with nitric acid at 70 °C, and a three dimensional lamellar structure clearly exists.

Fibrous polymer produced during the early stages of polymeriza-

I

II

III

IV

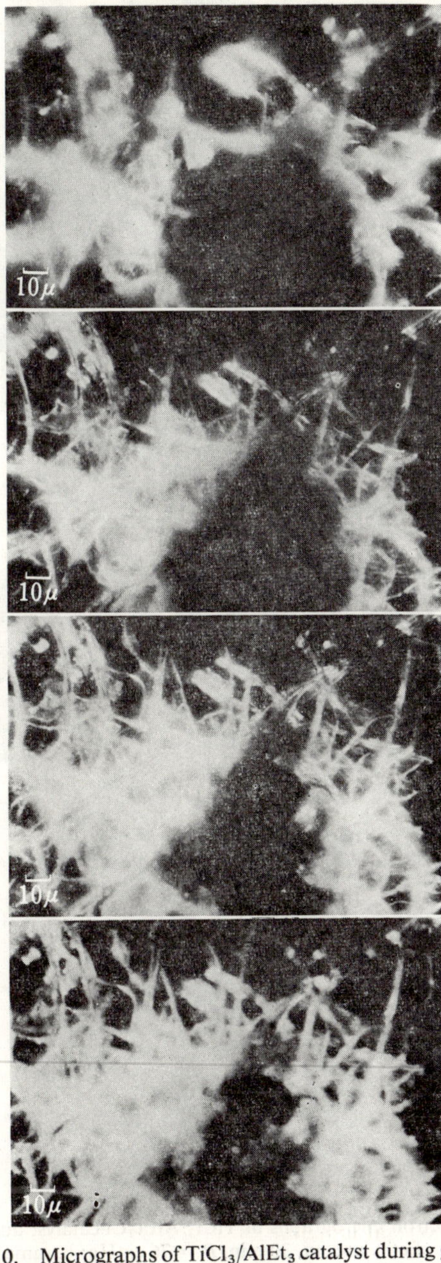

Fig. 7.10. Micrographs of $TiCl_3/AlEt_3$ catalyst during gas-phase polymerization of ethylene. (I, 5 min; II, 20 min; III, 30 min; IV, 45 min.)

a)

b)

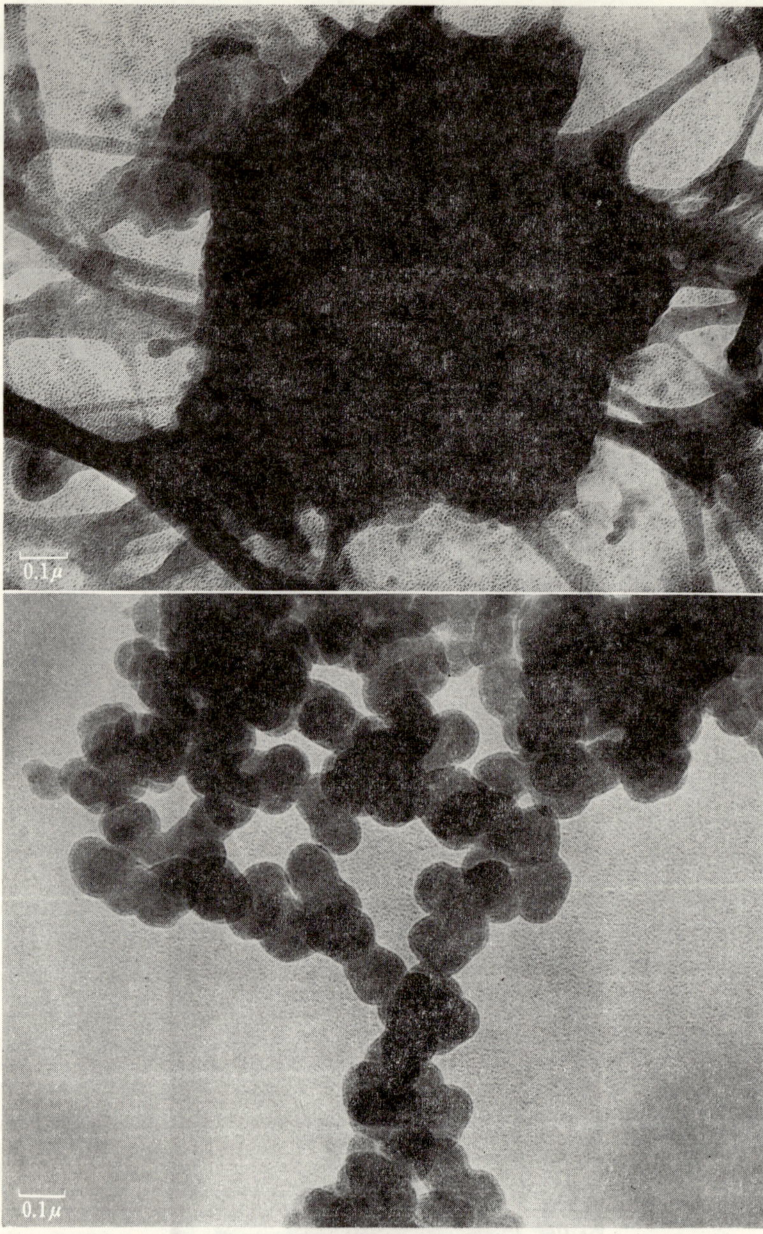

Fig. 7.11. Electron micrographs of $TiCl_3/AlEt_2Cl$ catalyst during gas-phase polymerization of ethylene. (a, 30 min, $P = 76$ cmHg; b, same system treated with nitric acid at 70°C for 24 hr after 15 min polymerization.)

tion was found to have a rather lower melting point (137 °C) than that produced later. The fibres do not appear to be extended crystals of molecular chains, but may be micelles of folded structures or clusters. From these results, it was concluded that two types of polymerization centres exist, one forming the lamellar structures and the other the fibrous structures.

The isotacticity of polypropylenes obtained at 10 atm and room temperature was rather higher than that found with the usual liquid-phase polymerization. The results are summarized in Table 7.4.

7.2 ALKYLATION OF AROMATIC SOLVENTS WITH $TiCl_3$

It is well known that $AlCl_3$ is a strong solid acid, and a good catalyst for the alkylation of some aromatic reagents, many of which are used as solvents in Ziegler-Natta polymerizations. $TiCl_3$ has also been found to possess strong acidity[2,5] as previously mentioned. It might therefore be expected that some solvent alkylation reactions should occur during Ziegler-Natta polymerization. Soga *et al.*[8] studied this problem, using AA-grade and TTC-grade $TiCl_3$. The latter catalyst is prepared from $TiCl_3$ by reduction with Ti metal, and is therefore free from $AlCl_3$. The following experiments were carried out: (1) polymerization of propylene at constant pressure (50 cm Hg) with $TiCl_3$ and $AlEt_3$ or $AlEt_2Cl$ using benzene or toluene solvent, (2) propylene,

TABLE 7.4 Gas-phase propylene polymerization†

$TiCl_3$ used (g)	$AlEt_3$ (as 10% solution) (cc)	Carrier	Polymerization time (min)	Polymers produced (g/g-$TiCl_3$)	Isotacticity (%)
0.330	5	none	18	100.5	84
0.385	5	polyethylene	60	45.6	
0.522	5	polyethylene	10	151	85
	$AlEt_2Cl$ (as 10% solution)				
0.248	5	polyethylene	25	21.2	92.5

† 25∼27 °C, 10 atm.

ethylene and acetylene at constant pressure (50 cm Hg) were each passed through benzene or toluene containing about 1 g-$TiCl_3$/250 ml of solvent, and (3) the above experiments were repeated at 300 cm Hg pressure in an autoclave with a magnetic stirrer. Typical curves showing adsorption rate *vs.* time for (1) and (2) are shown in Fig. 7.12.

The reaction products were quenched in 1 N HCl/methanol, and after separation of the polymers, the residual solution was analyzed by gas chromatography at 100°C, using a 1 m DLP-celite column with hydrogen as the carrier. The results are summarized in Table 7.5, and indicate that alkylation occurs in addition to polymerization even in the presence of alkylaluminium, and that in the absence of alkylaluminium, alkylation is the predominant reaction. Fig. 7.12 shows that such alkylations are not completely due to $AlCl_3$ in the $TiCl_3$ samples, since they also occured with TTC-grade $TiCl_3$. No essential difference can be seen between curve A and curves B, C and D, so that the rates of

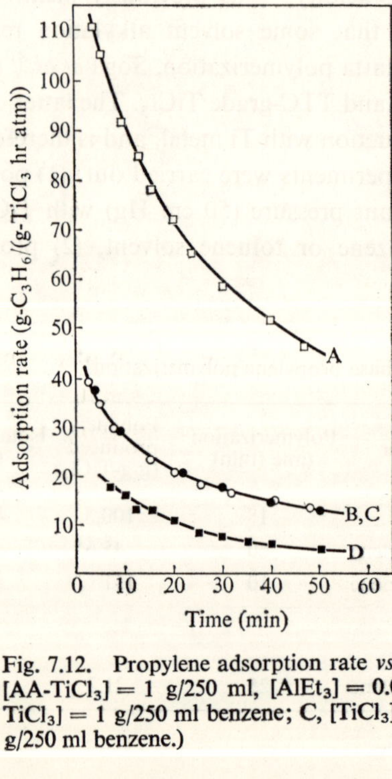

Fig. 7.12. Propylene adsorption rate *vs.* time. (P_{Pr} = 50 cmHg at 50°C. A, [AA-$TiCl_3$] = 1 g/250 ml, [$AlEt_3$] = 0.007 mole/250 ml benzene; B, [AA-$TiCl_3$] = 1 g/250 ml benzene; C, [$TiCl_3$] = 1 g/250 ml; D, [TTC-$TiCl_3$] = 1 g/250 ml benzene.)

TABLE 7.5 Variation of selectivity with various conditions

Solvent	Catalyst	Monomer	Pressure (cmHg)	Temp. (°C)	Product	Selectivity (%) $\left[100 \times \frac{\text{-R}}{\text{monomer absorbed}}\right]$ 
Benzene	$TiCl_3$ (AA)	C_3H_6	50	50	Cumene	~100
Benzene	$TiCl_3$(TTC)	C_3H_6	50	50	Cumene	~100
Toluene	$TiCl_3$(AA)	C_3H_6	50	50	Cymene	~100
Benzene	$TiCl_3$(AA)	C_2H_4	50	50	Ethylbenzene (EB)	~100
Benzene	$TiCl_3$(AA)	C_2H_2	40–65	30–65	—	—
Benzene	$TiCl_3$(AA)	C_2H_4	300	75	Polymer and EB	~50
Benzene	$TiCl_3$(AA)–$Al(C_2H_5)_2Cl$ (Al: Ti = 0.5)	C_3H_6	50	50	Polymer and cumene	~10
Benzene	$TiCl_3$(AA)–$Al(C_2H_5)_3$ (Al: Ti = 0.4)	C_3H_6	50	50	Polymer and cumene	~1
Benzene	$TiCl_3$(AA)–$Al(C_2H_5)_3$ (Al: Ti = 1)	C_3H_6	50	50	Polymer	0

polymerization and alkylation can scarcely be distinguished on the basis of the kinetic curves alone.

Fig. 7.13 shows the kinetic curve obtained on passing propylene into benzene containing $TiCl_3$, with subsequent addition of $AlEt_3$ (at the time indicated by the arrow) such that [A]/[Ti] becomes equal to unity. A comparision of curve A in Fig. 7.12 with curve B in Fig. 7.13 indicates that alkylation does not affect the formation of polymerization centres. However, the similarity between the alkylation and polymerization curves is quite marked, and the kinetic curve for alkylation (A) can in fact be represented by the equation

$$R = R_\infty(1 + fe^{-kt}) \tag{7.3}$$

which is of the same form as that governing decay type polymerization. Furthermore, the observed values of k were $1.5 \sim 1.6\ \text{hr}^{-1}$ at 35°C and $2.0 \sim 2.1\ \text{hr}^{-1}$ at 52°C, which compares with rate constants for polymerization of $1.5\ \text{hr}^{-1}$ and $2.0\ \text{hr}^{-1}$, respectively, at the same temperatures. It can therefore be concluded that the active sites for alkyla-

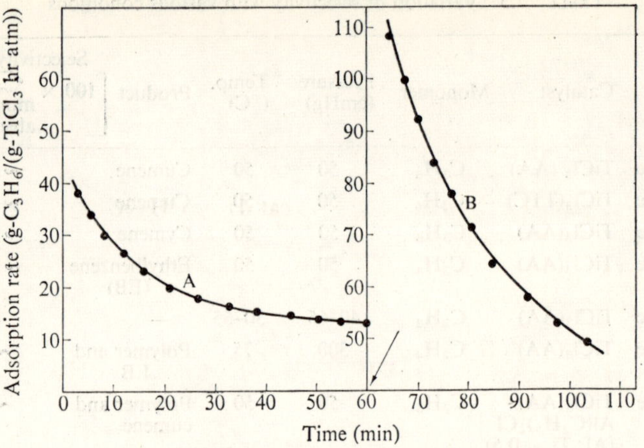

Fig. 7.13. Propylene adsorption rate *vs.* time. (A, P_{Pr} = 50 cmHg, [AA-$TiCl_3$] = 1 g/250 ml benzene at 50°C. B, $AlEt_3$ added at the point indicated by the arrow. [A]/[Ti] = 1.)

tion and for polymerization are different, but that they are deactivated by the same mechanism.

7.3 HYDROGEN EXCHANGE AND HYDROGENATION WITH ZIEGLER-NATTA CATALYSTS

Ziegler-Natta systems are known to catalyze hydrogen/deuterium exchange reactions. Schindler[9,10] has studied ethylene polymerization with $TiCl_4$ or $TiCl_3$ and $AlEt_3$ or $AlEt_2Cl$, and observed that exchange occurred between gaseous hydrogen and hydrogen on the β-carbon of the growing polymer chain. Schindler also noted[11] the formation of deuterated ethanes by $TiCl_4/AlEt_3$ at high [A]/[Ti] ratios. The hydrogenation of olefins by soluble complexes such as $Ti(O\text{-}i\text{-}C_3H_7)_4$, $(C_5H_5)_2TiCl_2$-$(i\text{-}C_4H_9)_3Al$, $AlEt_3$ has also been observed by Sloan *et al.*[12]

From his comprehensive studies, Schindler[13] concluded that the following mechanisms accounted for the experimental data. In the presence of Ti (IV) sites at low [A]/[$TiCl_4$] catalyst ratios, an exchange reaction between adsorbed deuterium and the hydrogen on the β-carbon of the growing polymer chain is favoured.

$$\begin{matrix} \text{cat–CH}_2\text{–CH}_2\text{–R} \\ \text{D–D} \end{matrix} \rightleftharpoons \begin{matrix} \text{cat–CH}_2\text{–CHD–R} \\ \text{H–D} \end{matrix} \rightleftharpoons \begin{matrix} \text{cat–CH}_2\text{–CD}_2\text{–R} \\ \text{H–H} \end{matrix} \qquad (7.4)$$

With Ti (III) sites, however, only a transfer step will occur

$$\text{cat–CH}_2\text{–CH}_2\text{–R} + \text{H}_2 \longrightarrow \text{cat–H} + \text{CH}_3\text{–CH}_2\text{–R} \qquad (7.5)$$

At high catalyst ratios, Ti (II) sites become important, giving rise to the following reactions

$$\text{cat–H} \xrightarrow{\text{CH}_2=\text{CH}_2} \text{cat–CH}_2\text{–CH}_3 \xrightarrow{\text{D}_2} \text{cat–D} + \text{CH}_3\text{–CH}_2\text{D} \qquad (7.6)$$

$$\text{cat–D} + \text{CH}_2=\text{CH}_2 \rightleftharpoons \text{cat–CH}_2\text{–CH}_2\text{D} \rightleftharpoons \text{cat–H} + \text{CH}_2=\text{CHD} \qquad (7.7)$$

which can account for the formation of all types of deuteroethylene and deuteroethane up to C_2D_6. Schindler also considered the possibility of *cis-trans* isomerization, or double bond migration by the scheme shown in Eq. 7.8

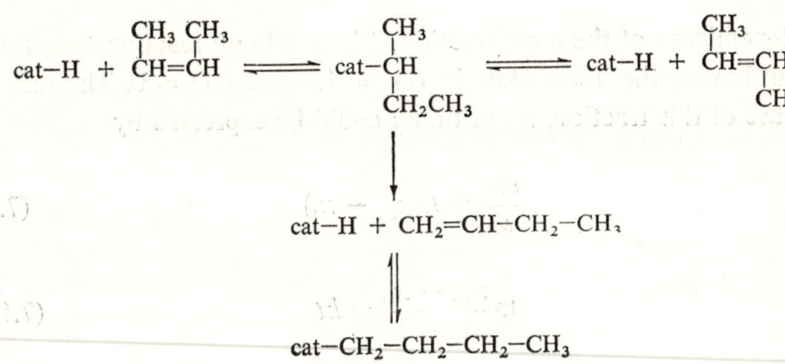

(7.8)

7.3.1 Kinetics of H_2/D_2 exchange with $TiCl_3$/alkylaluminium

Kojima *et al.*[14)] have investigated H_2/D_2 exchange at low pressures (10 cm Hg) using a 100 ml flask with a magnetic stirrer, and 5 ml of *n*-heptane as solvent. The catalysts used were $TiCl_3$ (AA-grade, 20 m^2/g, and TAC-grade) with $AlEt_3$, $AlEt_2Cl$, $AlEtCl_2$, $AlMe_3$, $Al(i\text{-}Bu)_3$, or $ZnEt_2$. Isotopic analysis of the products was carried out by low pressure gas chromatography and mass spectroscopy. Fig. 7.14 shows

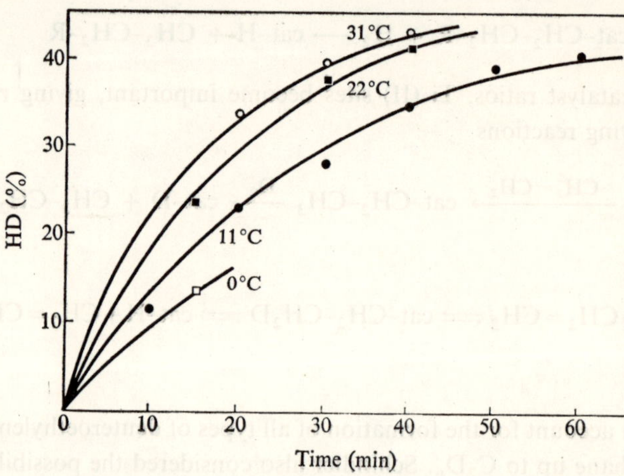

Fig. 7.14. Mole fraction of HD in gas phase *vs.* reaction time at various temperatures. ($TiCl_3$ = 0.2 g, $AlEt_3$ = 0.6 mmole, P = 9.6 cmHg.)

the dependence of the mole fraction (%) of HD on reaction time for a system having the initial state $H_2/D_2 = 1.0$ and HD = 0. The rate of increase of this fraction, x_t, at time t could be expressed by

$$\frac{dx_t}{dt} = k(x_{eq} - x_t) \tag{7.9}$$

$$\ln \frac{x_{ea} - x_0}{x_{eq} - x_t} = kt \tag{7.10}$$

where x_{eq} is the fraction of HD at equilibrium, and x_0 is the fraction at $t = 0$ ($x_0 = 0$ in this experiment). The applicability of Eq. 7.10 to the experimental results given in Fig. 7.14 is shown in Fig. 7.15. The dependence of k on the amount of $TiCl_3$ used and on the total pressure is shown in Fig. 7.16 and 7.17. Fig. 7.18 shows the temperature dependence of k (1/(g-$TiCl_3$ min)).

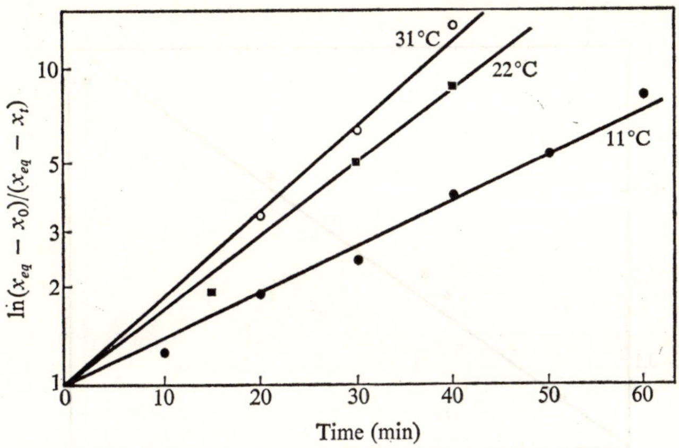

Fig. 7.15. Applicability of Eq. 7.10 to the experimental data of Fig. 7.14.

The same reaction has been studied using $TiCl_3/AlEt_2Cl$, and though the absolute value of k is smaller, the results are very similar to those discussed above (Fig. 7.19 and 7.20). The results obtained with other catalyst systems are summarized in Table 7.6.

The activities of these catalysts in the H_2/D_2 exchange reaction are proportional to the reducing power of the cocatalyst

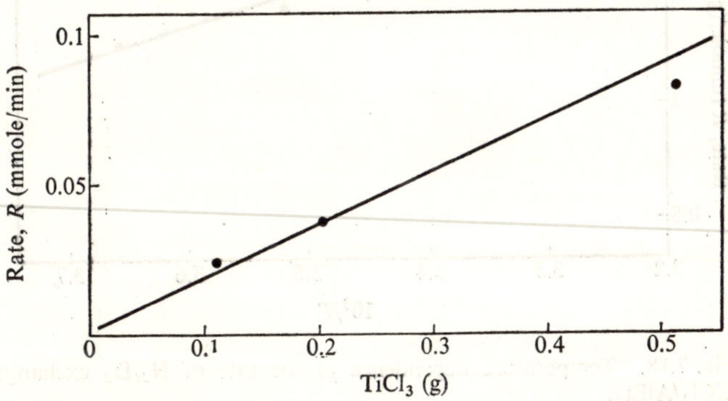

Fig. 7.16. Dependence of rate of H_2/D_2 exchange on mass of $TiCl_3$. ($AlEt_3$ = 0.6 mmole at 11°C. P = 9.6 cmHg.)

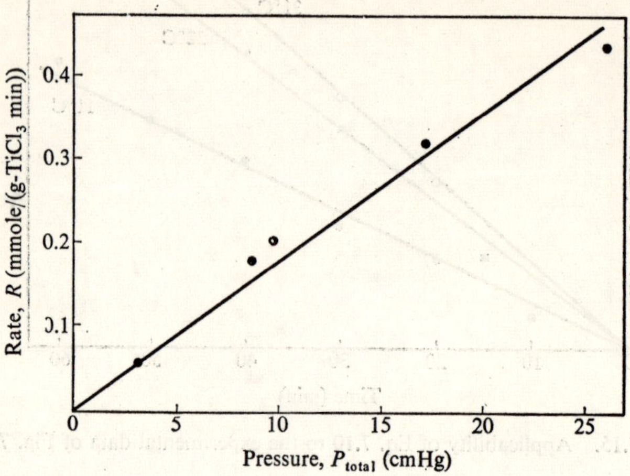

Fig. 7.17. Pressure dependence of the rate of H_2/D_2 exchange. ($TiCl_3$ = 0.2 g, $AlEt_3$ = 0.6 mmole at 11°C.)

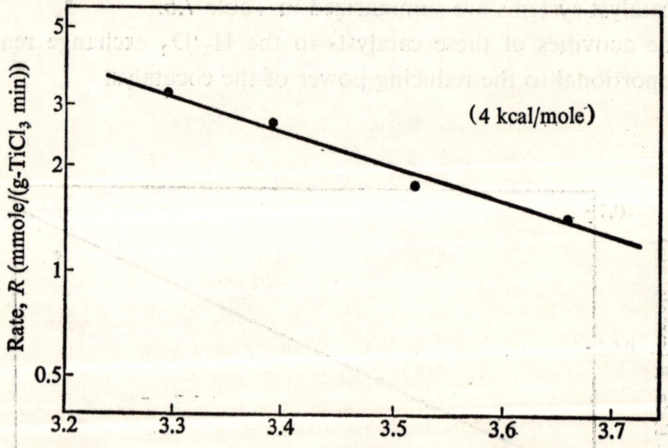

Fig. 7.18. Temperature dependence of the rate of H_2/D_2 exchange with $TiCl_3/AlEt_3$.

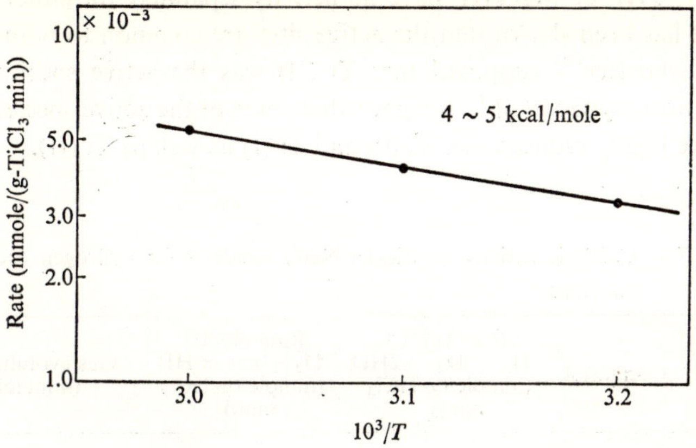

Fig. 7.19. Temperature dependence of the rate of H_2/D_2 exchange with $TiCl_3/AlEt_2Cl$ ($TiCl_3$ = 1.2 g, $AlEt_2Cl$ = 3 mmole, P = 19.2 cmHg.)

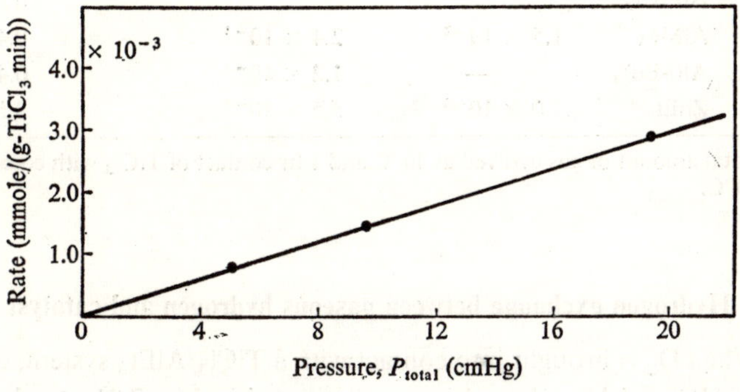

Fig. 7.20. Pressure dependence of the H_2/D_2 exchange rate. ($AlEt_2Cl$ = 3 mmole at 40°C.)

$$AlEt_3 > AlMe_3 > ZnEt_2 > AlEt_2Cl \gg AlEtCl_2 = 0 \qquad (7.11)$$

The amount of gas evolved on contact of the alkyl aluminium with the $TiCl_3$ varies as follows

$$ZnEt_2 > AlEt_3 > AlMe_3 > AlEt_2Cl \qquad (7.12)$$

It is interesting to note that the $TiCl_3$/THF/Mg system is active in hydrogen exchange reactions, as well as in nitrogen fixation.[15] Mixtures of $TiCl_3$/THF or Mg/THF have no activity separately for either reaction. It has been shown that the active sites are common to both reactions. Schindler[13)] supposed that Ti (II) was the active species, but recent work by Go *et al.*[15)] suggests that some of the active species may be more highly reduced, i.e. Ti (0) and Ti (I) as well as Ti (II).

TABLE 7.6 Catalytic activity of Ziegler-Natta catalysts for hydrogen exchange reactions

Catalyst	Cocatalyst	Rate (11 °C) $H_2 + D_2 \rightleftharpoons 2HD$ (mmole/(g-$TiCl_3$ min))	Rate (40 °C) D_2 + cat $\rightleftharpoons$ HD (mmole/(g-$TiCl_3$ min))	Gas evolution†1 (mmole)
$TiCl_3$	—			0
—	$AlEt_3$			0
$TiCl_3$	$AlEt_3$	2.1×10^{-1}	1.2×10^{-2}	1.1
$TiCl_3$	$AlEt_2Cl$	1.5×10^{-3} †2	negligible	0.2
$TiCl_3$	$AlEtCl_2$			0
$TiCl_3$	$AlMe_3$	1.5×10^{-2}	2.4×10^{-3}	0.49
$TiCl_3$	Al(*i*-Bu)$_3$	—	1.2×10^{-2}	0.46
$TiCl_3$	$ZnEt_2$	7.0×10^{-3}	5.5×10^{-3}	3.56

†1 Total amount of gas evolved at 40 °C and 1 hr contact of $TiCl_3$ with cocatalyst.
†2 40 °C.

7.3.2 Hydrogen exchange between gaseous hydrogen and catalyst

When D_2 is brought into contact with a $TiCl_3/AlEt_3$ system, evolution of HD and H_2 takes place, as illustrated in Fig. 7.21. Analysis of the gas phase (Fig. 7.22) shows that equilibrium between D_2, HD and H_2 is established before equilibrium between the various gases and the catalyst system. It has been confirmed that consumption of hydrogen ceases after about one hour. The consumption of hydrogen in the system was about 0.15 mmole, corresponding to monolayer formation (10^{15} molecules/cm^2) on the surface of the $TiCl_3$ (TAC-grade, specific surface area 20 m^2/g). The exchangeable hydrogen atoms in the system were equivalent to 4/5 of the total moles of $AlEt_3$.

During hydrogen exchange between the gas and the catalyst system, a considerable amount of ethane was formed. Evolution of ethane reached its maximum value after two hours, while the volume

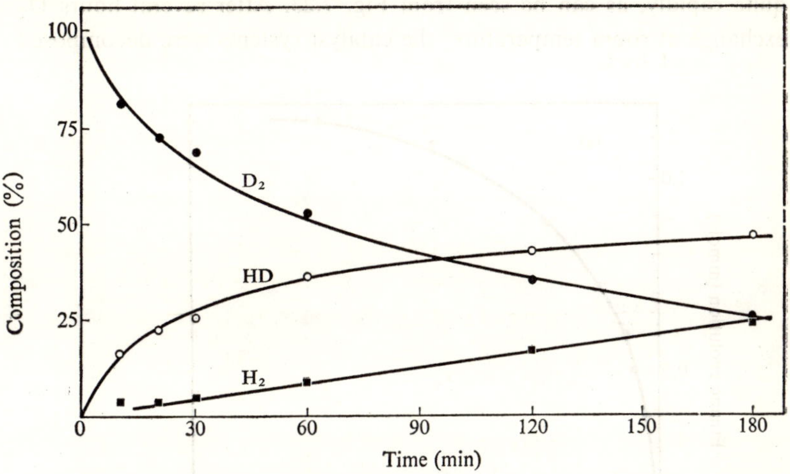

Fig. 7.21. Changes in the composition of the gas-phase during the exchange reaction between D_2 and the catalyst system. ($TiCl_3$ = 1.10 g, $AlEt_3$ = 3 mmole, initial pressure P_D = 10.6 cmHg.)

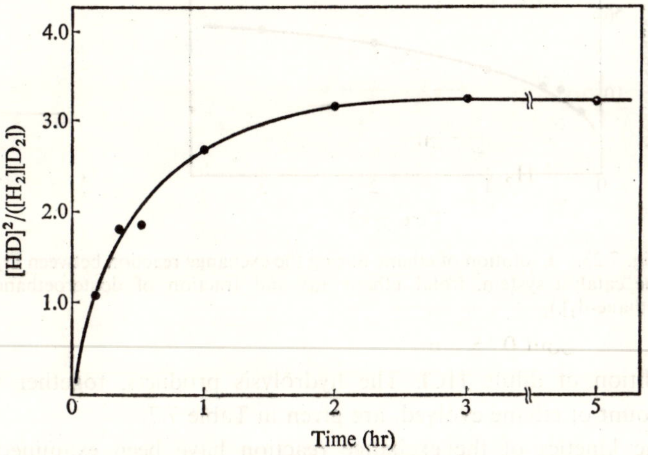

Fig. 7.22. Establishment of exchange equilibrium between D_2 and the catalyst system. The equilibrium constant is 3.29 at 40°C.

percent of deutero-ethane (mainly ethane-d_1) reached a constant value quite rapidly, as can be seen from Fig. 7.23. After several hours D_2 exchange at room temperature, the catalyst systems were decomposed

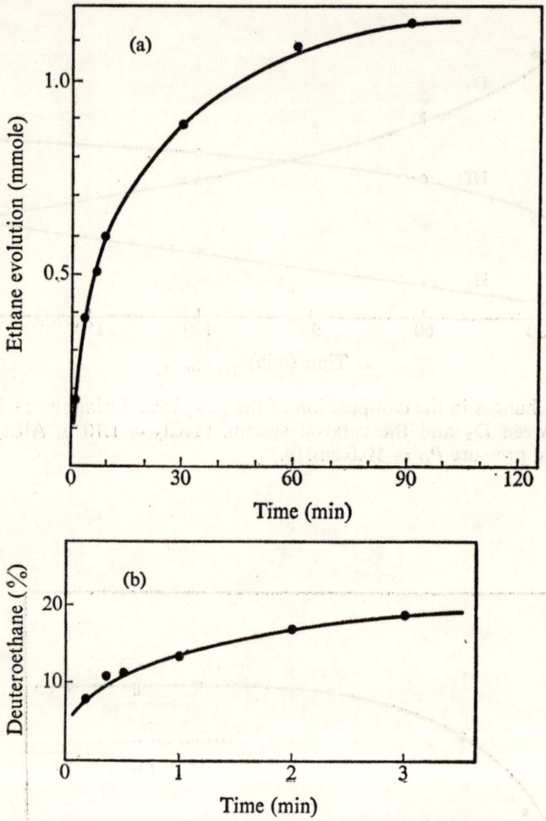

Fig. 7.23. Evolution of ethane during the exchange reaction between D_2 and the catalyst system. (total ethane (a) and fraction of deuteroethane (b) [ethane-d_1].)

by addition of dilute HCl. The hydrolysis products, together with the amount of ethane evolved, are given in Table 7.7.

The kinetics of the exchange reaction have been examined by Kojima and Keii,[14)] who found that the initial increasing rate, R_0, of formation of H_2 and HD is proportional to $(P_{D_2})^{1/2}$, and is independent of $AlEt_3$ concentration over the range 0.1 ~ 0.6 mole/l. The activation

TABLE 7.7 Results of hydrolysis of $TiCl_3$-$AlEt_3$ after hydrogen exchange with D_2†1

	$AlEt_3$ = 3 mmole D_2 = 1.7 mmole	$AlEt_3$ = 3 mmole D_2 = 1.0 mmole
Gas ($2H_2$ + HD)	1.3 mmole	1.0 mmole
Ethane evolved during exchange	1.5 (0.3)†2 mmole	1.7 (0.3)†2 mmole
Ethane evolved by hydrolysis	6.0 (0.6)†2 mmole	6.6 (0.4)†2 mmole
Polyethylene†3 (as ethylene)	1.5 mmole	0.7 mmole

†1 Room temperature, 5 hr, $TiCl_3$ = 1.2 g.
†2 Figures in parentheses indicate deuteroethane (mmole).
†3 Calculated from the material balance, $3[AlEt_3] = C_2H_6$ evolved + ethylene polymerized.

energy was found to be 6 kcal/mole. The above results are illustrated in Fig. 7.24, 7.25 and 7.26.

The initial rate, R_0, can be represented at higher $AlEt_3$ concentrations by

$$R_0 = k_0 e^{-6\text{kcal}/RT} P_{D_2}^{1/2} [A]_0 \tag{7.13}$$

Similar experiments were carried out with other catalyst systems consisting of $TiCl_3$ and $AlEt_2Cl$, $AlMe_3$, $Al(i\text{-}Bu)_3$ or $ZnEt_2$, and the results are summarized in Table 7.6. From the rate law given in Eq. 7.13, it appears that the rate determining step in the exchange reaction may be the reaction between an adsorbed deuterium atom and a hydrogen atom on the β-carbon of an $AlEt_3$ ethyl group. The reaction scheme may therefore be written as

a) $$D_2\ (\text{gas}) \underset{k'_H}{\overset{k_H}{\rightleftharpoons}} 2D\ (a)$$

b) $$Al_2Et_6 \underset{k'_A}{\overset{k_A}{\rightleftharpoons}} Al_2Et_6\ (a) \tag{7.14}$$

c) $$D(a) + Al_2Et_6\,(a) \xrightarrow{k_e} HD + Al_2Et_5(C_2H_4D)(a)$$

It was observed that the exchange rate was strongly depressed on removal of the liquid phase, indicating that interchange of $AlEt_3$

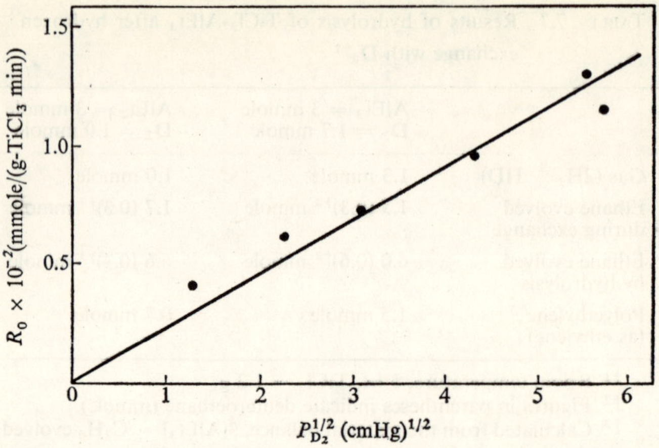

Fig. 7.24. Pressure dependence of the initial rate of exchange between D_2 and the catalyst system. ($TiCl_3$ = 0.55 g $AlEt_3$ = 3 mmole at 11°C.)

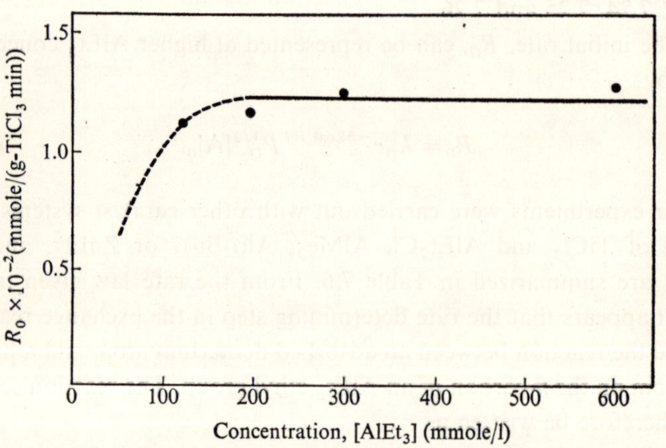

Fig. 7.25. Dependence of the initial exchange rate on $AlEt_3$ concentration ($TiCl_3$ = 0.55 g at 22°C. P_D = 9.6 cmHg.)

between the catalyst surface and the liquid phase is fast, which is in agreement with the results of Takegami.[16] Evolution of ethane may occur by the following processes

d) $$Al_2Et_6(a) \xrightarrow{k_d} Al_2Et_4(a) + C_2H_6 + C_2H_4(a)$$

d′) $$AlEt_3(a) \longrightarrow AlEt(a) + C_2H_6 + C_2H_4(a)$$

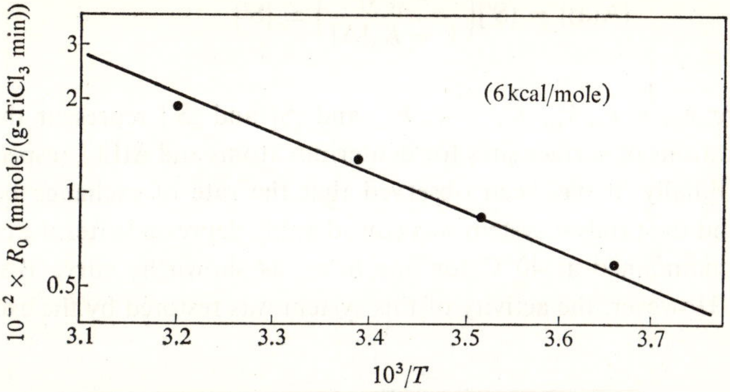

Fig. 7.26. Temperature dependence of initial rate of exchange between D_2 and catalyst system.

where the ethylene produced is polymerized on the catalyst surface. Since the fraction of ethane-d$_1$ is small, the evolution of ethane can be considered to be mainly due to spontaneous processes such as (d) or (d′). In this case, formation of ethane-d$_1$ can be considered as due to the decomposition of deuterated $AlEt_3$ on the catalyst surface. The deuterium content of evolved ethane was higher than that of the ethane produced on hydrolysis, from which we may conclude that

$$k_d > k'_A \gg k_e \tag{7.15}$$

and hence that

$$\frac{d[\text{HD}]}{dt} = \frac{d[\text{ethane-d}_1]}{dt} \tag{7.16}$$

Fig. 7.23 and 7.24 show that the amount of ethane-d$_1$ evolved in the first 10 min is 0.12 mmole (= 0.6 mmole × 0.2) and the corresponding amount of HD is 0.12 mmole.

The rate law given in Eq. 7.13 can be derived from the above scheme provided that $K_H P_{D_2} \ll 1$ and $K_A[\text{A}] \gg 1$ for the adsorptions of deuterium and $AlEt_3$ on independent sites. Thus we have

$$[\text{D(a)}] = [\text{S}]\left(\frac{(K_H P_{D_2})^{1/2}}{1 + (K_H P_{D_2})^{1/2}}\right) \simeq [\text{S}] K_H^{1/2} P_{D2}^{1/2}$$

$$[A(a)] = [S']\left(\frac{K_A[A]}{1 + K_A[A]}\right) \simeq [S']$$

where $K_H = k_H/k'_H$, $K_A = k_A/k'_A$, and [S] and [S′] represent the concentrations of surface sites for deuterium atoms and $AlEt_3$, respectively.

Finally, it has been observed that the rate of exchange between D_2 and the catalyst system was considerably depressed after the catalyst was maintained at 40 °C for one hour, as shown by curve B in Fig. 7.27. However, the activity of this system was restored by the exchange

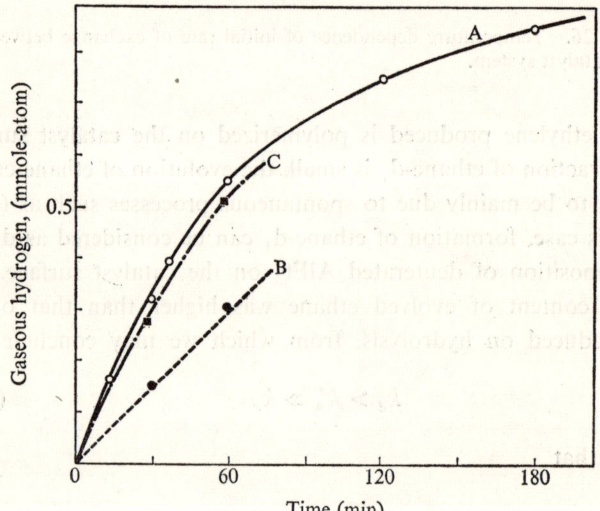

Fig. 7.27. Effect of catalyst ageing on the rate of exchange between D_2 and catalyst. ($TiCl_3$ = 0.55 g, $AlEt_3$ = 3 mmole at 40°C. P_D = 9.6 cmHg. A, aged at room temperature for 10 min; B, aged at 40°C for 1 hr (evolved ethane = 1.2 mmole); C, aged at 40°C for 1 hr. $AlEt_3$ removed and fresh $AlEt_3$ substituted.)

of the liquid phase for a fresh solution of $AlEt_3$. Evolution of ethane (1.2 mmole) was observed during the ageing of the catalyst. It may therefore be considered that the loss of a small fraction of the available ethyl groups, 1.2 mmole in 3 × 3 mmole, was responsible for a reduction of about 50% in catalyst activity. This suggests that one ethyl group in $AlEt_3$ is more active than the others in exchange reactions.

7.4 RELATED ELEMENTARY REACTIONS OF HOMOGENEOUS CATALYSTS

A number of soluble complexes based on transition metals are known to act as catalysts similar in some respects to Ziegler-Natta catalysts. Since the structure of these complexes is generally well established, their reaction mechanisms may cast some light on the nature of the active centre in Ziegler-Natta systems.

Shilov[17] reviewed the role of mono- and binuclear complexes in homogeneous catalysis, and summarized the important basic reactions as follows

Addition

$$\rangle M + H_2 \rightleftharpoons \rangle M\langle^{H}_{H} \rightleftharpoons [\rangle MH]^{\ominus} + H^{\oplus} \quad (7.17)$$

$$\rangle M + RH_2 \rightleftharpoons \rangle M\langle^{R}_{H} \rightleftharpoons [\rangle MR]^{\ominus} + H^{\oplus} \quad (7.18)$$

Insertion

$$\rangle MH + \rangle C{=}C\langle \rightleftharpoons \begin{array}{c}\rangle M{-}H\\ |\\ \rangle C{=}C\end{array} \rightleftharpoons \rangle M{-}\overset{|}{\underset{|}{C}}{-}\overset{|}{\underset{|}{C}}{-}H \quad (7.19)$$

$$\rangle M{-}R + \rangle C{=}C\langle \rightleftharpoons \rangle M{-}\overset{|}{\underset{|}{C}}{-}\overset{|}{\underset{|}{C}}{-}R$$

$$\rangle M{-}R + \rangle C{=}O \rightleftharpoons \rangle M{-}COR \quad (7.20)$$

Substitution

$$\rangle M{-}R + H_2 \rightleftharpoons \rangle M{-}H + RH \quad (7.21)$$

$$\rangle M{-}R + H_2O \rightleftharpoons \rangle M{-}OH + RH \quad (7.22)$$

The insertion of olefin into M–H bonds has been observed by Chatt *et al.*[18] and Wilkinson *et al.*[19] for Pt and Co complexes. Insertion into an M–R bond has also been confirmed spectroscopically by Shilov *et al.*[20,21] in the following reaction.

$$\begin{array}{c}(C_5H_5)_2TiCH_3\\ |\\ Cl{-}AlRCl_2\end{array} + C_2H_4 \longrightarrow \begin{array}{c}(C_5H_5)_2Ti(C_3H_7)\\ |\\ Cl{-}AlRCl_2\end{array} \quad (7.23)$$

Shilov considered that the insertion of ethylene only took place with the species $(C_5H_5)_2TiR^+$. The catalytic dimerization of ethylene (insertion of ethylene into an M–R bond) by $RhCl_3$ was studied spectroscopically by Cramer[22] who put forward the mechanism given in Eq. 7.24.

$$
\begin{aligned}
&[C_2H_5Rh^{III}Cl_2(solv)]^{\ominus} + C_2H_4 \rightleftharpoons [C_2H_5Rh^{III}Cl(C_2H_4)(solv)]^{\ominus} \\
&\xrightarrow{C_2H_4} [CH_3CH_2CH_2CH_2Rh^{III}Cl_3(solv)]^{\ominus} \\
&\underset{+HCl}{\overset{-HCl}{\rightleftharpoons}} [Cl_2Rh^{I}(solv)(CH_3CH_2CH_2CH_2)] \\
&\xrightarrow{C_2H_4} [Cl_2Rh^{I}(C_2H_4)_2]^{\ominus} + \text{butene-1} \qquad (7.24)
\end{aligned}
$$

Ketley *et al.*[23] examined the dimerization of ethylene with $PdCl_2$ in CH_2Cl_2 or $C_6H_5 \cdot NO_2$, and were able to isolate the various reaction intermediates. The proposed mechanism is

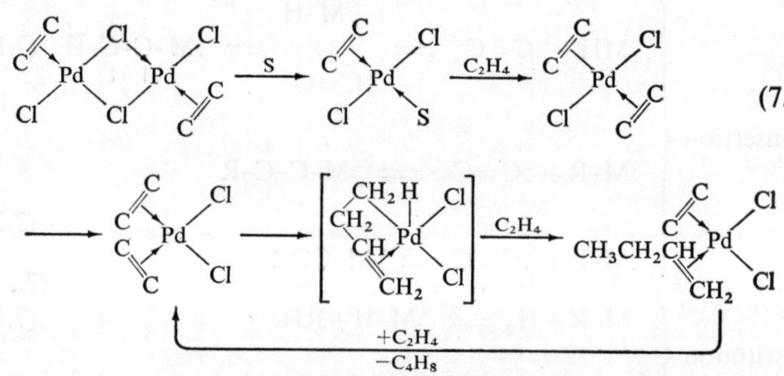

(7.25)

where S is a solvent molecule. It is important to note that this mechanism involves the cleaveage of the binuclear complex by a solvent molecule to form a mononuclear complex, the *cis-trans* isomerization of $PdCl_2(C_2H_4)_2$, and the liberation of butene due to coordination of fresh C_2H_4.

Systematic studies have been made by Yamamoto *et al.*[24] on the interactions between olefins and transition metal alkyl dipyridyl complexes such as R_2Ni (dipy) and R_2Fe $(dipy)_2$. The structures of some of these complexes are given below

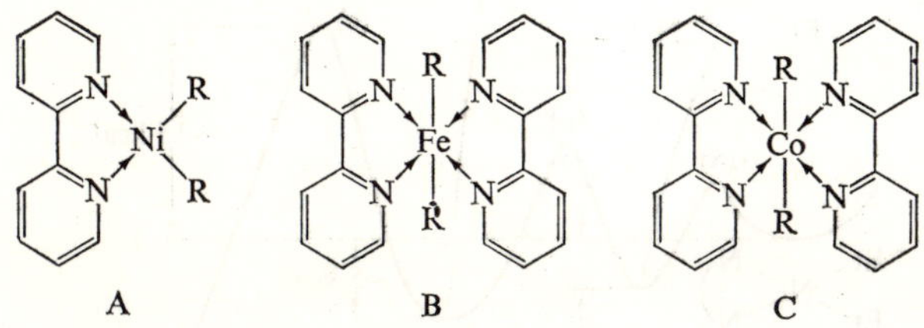

(7.26)

These complexes are quite stable, but decompose thermally to form R–R. In the presence of olefins, high polymers are formed. The stability of the R-metal bond has been found to depend on the nature of R such that *i*-Bu < Pr < Et < Me. Thus complex A with R = *i*-Bu decomposes at below 0 °C, R = Pr decomposes gradually in the range room temperature ~ 110 °C (16 kcal/mole activation energy), R = Et decomposes above 110 °C (68 kcal/mole), and when R = Me the complex is stable up to 160 °C.

The reactions of these complexes with olefins may well have some bearing on Ziegler-Natta catalysis. The liberation of R–R during the reaction with olefin is shown in Eq. 7.27.

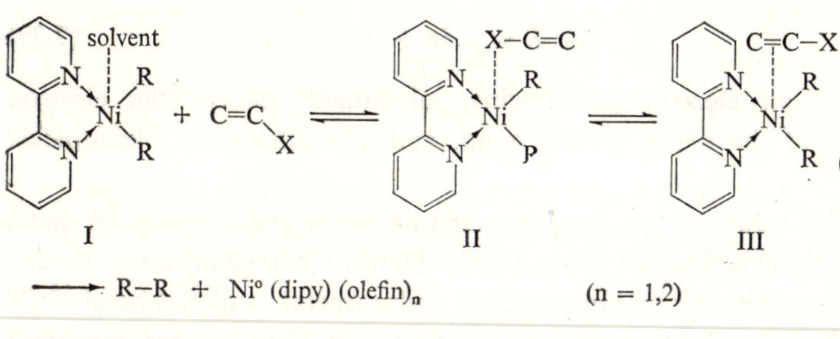

(7.27)

The potential energy reaction coordinate for Eq. 7.27 is illustrated in Fig. 7.28. The activation energy for butane liberation is about 15 kcal/mole, whereas that for the pyrolysis of I is about 68 kcal/mole. The coordination of olefin is responsible for this lowering of the activation energy. Yamamoto *et al.* obtained values of the equilibrium constant, *K*, of the reaction

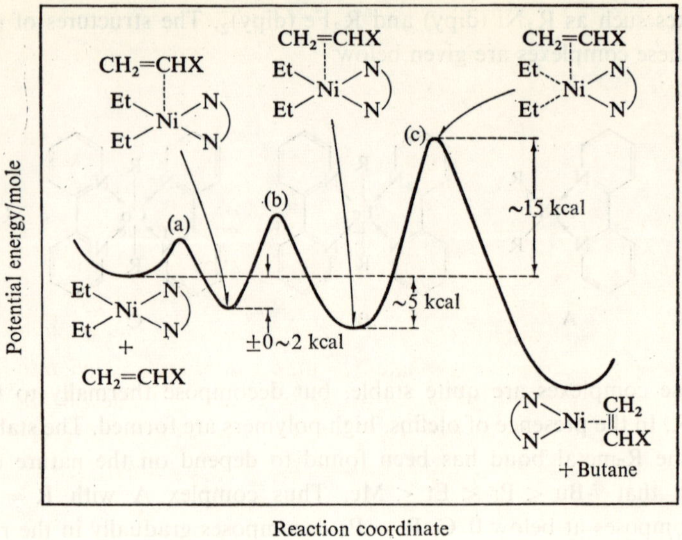

Fig. 7.28. Reaction of Et_2Ni (dipy) with olefin.[20]

$$\mathrm{Ni(dipy)(solv)} + \mathrm{olefin} \overset{K}{\rightleftharpoons} \mathrm{Ni(dipy)(olefin)} + \mathrm{solvent} \quad (7.28)$$

for a wide range of olefins (e.g. vinyl acetate, methyl acrylate, methyl vinyl ketone, acrolein, etc.), and found the following relation between log K at 30°C and the Alfrey-Price e-values[25]

$$\log K = 2.3e + 0.3 \quad (7.29)$$

Clearly, an electron-withdrawing substituent reduces the electron density at the olefin double bond, and strengthens the olefin/nickel bond.

The nature of the bond between an olefin and a transition metal is usually explained in terms of the Dewar-Chatt-Duncanson model, with a combined σ (olefin → metal) and π (metal → olefin) bond.[18,26] The above results therefore suggest that π-bonding is more important for the formation of a stable complex. The π-bond is formed by the overlap of a filled metal d-orbital with the anti-bonding π^*-orbital of the olefin, and will therefore be strongest when the original energy levels of the d- and π^*-orbitals are similar.[27] This is compatible with the observation of a linear relation between log K and the energies of the lowest vacant π^* orbitals of various olefins (calculated by Kagiya *et al.*[28]). Yamamoto *et al.* also found a linear relationship between the equilibrium constant, K, of Eq. 7.28 and the rate constant of the corresponding reaction (Eq. 7.27)

$$\log k = 1.4 \log K - 4.7 \tag{7.30}$$

from which it was concluded that the coordination of olefin to nickel was responsible for the activation of the Ni-R bond.

The reactions of dialkylbis(pyridyl) iron, B, with various substituted olefins were also studied by Yamamoto *et al.* Some olefins such as acrylamide gave coordinated complexes, whereas vinyl monomers with large *e*-values were polymerized, as shown in Table 7.8. The reaction of B with these olefins involves partial dissociation of the dipyridyl ligand, followed by coordination of the olefin, and activation of the Fe–R bond.

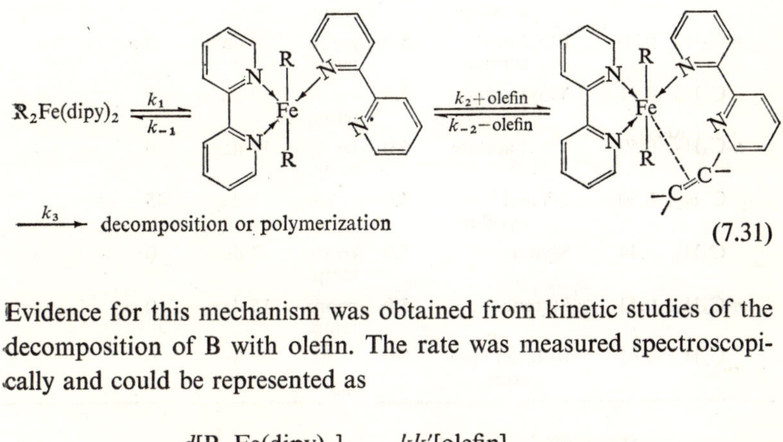

(7.31)

Evidence for this mechanism was obtained from kinetic studies of the decomposition of B with olefin. The rate was measured spectroscopically and could be represented as

$$r = -\frac{d[R_2Fe(dipy)_2]}{dt} = \frac{kk'[\text{olefin}]}{k + k'[\text{olefin}]}[R_2Fe(dipy)_2] \tag{7.32}$$

In terms of the scheme in Eq. 7.31, $k = k_1$ and $k' = k_3K_1K_2$. It was confirmed that the rate constant, k_1, decreased with R in the order n-Pr > Et > Me, and that k' correlated with K (of Eq. 7.28) such that $\log k' \propto \log K$. These results again suggest the importance of the coordination of olefin to metal in the activation of the metal-alkyl bond, prior to its fission, or to insertion of the olefin.

TABLE 7.8 Polymerization of vinyl monomer by $R_2Fe(dipy)_2$[20]

$R_2Fe(dipy)_2$ (g)		Monomer	(ml)	Temp. (°C)	Time	Yield (%)	$M \times 10^{-4}$
R = C_2H_5,	0.050	Acrylonitrile	5.0	30	9 min	30	1.05
C_2H_5,	0.020	Methacrylonitrile	10.0	room temp.	3 days	53	12.7
C_2H_5,	1.04	Acrolein	6.0	−30	2 days	70	
C_2H_5	1.37	Methyl vinyl ketone	5.0	−20	10 days	77	
C_2H_5,	0.24	Methyl acrylate	6.0	room temp.	18 hr	64	
CH_3,	0.040	Methyl methacrylate	10.0	25	6 days	16	
C_2H_5,	0.13	Methyl methacrylate	10.0	31	3 days	5.6	12.6
n-C_3H_7,	0.38	Methyl methacrylate	8.0	30	2 days	33	
CH_3,	0.35	Ethyl methacrylate	10.0	room temp.	2 days	trace	
C_2H_5,	0.23	Ethyl methacrylate	10.0	room temp.	7 days	0	
n-C_3H_7,	0.20	Ethyl methacrylate	5.0	room temp.	9 days	3.6	
C_2H_5,	0.91	Vinyl acetate	5.5	room temp.	3 days	0	
n-C_3H_7,	0.50	Vinyl acetate	6.0	room temp.	17 days	0	
C_2H_5,	0.30	2-Vinyl pyridine	4.0	20	3 days	45	
C_2H_5,	0.44	Styrene	5.0	room temp.	7 days	0	
n-C_3H_7,	0.43	Styrene	5.0	room temp.	15 days	0	
C_2H_5,	0.62	Isobutyl vinyl ether	5.0	50	20 hr	0	

REFERENCES

1. T. Keii, T. Takagi and S. Kanetaka, *Shokubai* (Japanese), **3**, 210 (1961); *Kogyo Kagaku Zasshi* (Japanese), **67**, 1433 (1964).
2. K. Tanabe and Y. Watanabe, *Shokubai* (Japanese), **4**, 360 (1962); *J. Res. Inst. Catalysis, Hokkaido Univ.*, **11**, 65 (1963); K. Tanabe, *Solid Acids and Bases*, p. 92, Kodansha-Academic Press, 1970.

3. S. Kanetaka, T. Takagi and T. Keii, *Kogyo Kagaku Zasshi* (Japanese), **67**, 1436 (1964).
4. K. Soga and T. Keii, unpublished paper.
5. Y. Doi, A. Yoshimoto and T. Keii, presented at Annual Meeting of Japan Chem. Soc., 1971.
6. S. Kanetaka and T. Takagi, *Kogyo Kagaku Zasshi* (Japanese), **69**, 695 (1966); S. Kanetaka, *Kogyo Kagaku Zasshi* (Japanese), **70**, 1568 (1967).
7. I. Okura, T. Matsumoto, S. Kanetaka, T. Kawai, H. Maeda and T. Keii, presented at Meeting of Japan Cellulose Soc., 1968.
8. K. Soga, T. Keii and A. Takahashi, *J. Polymer Sci.*, **B3**, 1075 (1965).
9. A. Schindler, *J. Polymer Sci.*, **C4**, 81 (1964).
10. A. Schindler, *J. Polymer Sci.*, **B3**, 793 (1964).
11. A. Schindler, *Makromol. Chem.*, **118**, 1 (1968).
12. M.F. Sloan, A.S. Matlack and D.S. Breslow, *J. Amer. Chem. Soc.*, **85**, 4014 (1963).
13. L. Reich and A. Schindler, *Polymerization by Organometallic Compounds*, p. 315, Interscience, 1966.
14. A. Kojima and T. Keii, unpublished paper; A. Kojima, *M. Eng. Thesis, Tokyo Inst. Technology*, 1968.
15. S. Go and T. Keii, unpublished paper; S. Go, *M. Eng. Thesis, Tokyo Inst. Technology*, 1968.
16. Y. Takegami and T. Suzuki, *Shokubai* (Japanese), **7**, 11, 1965.
17. A.E. Shilov, *Kinetika i Kataliz*, **11**, 305 (1970).
18. J. Chatt and L.A. Duncanson, *J. Chem. Soc.*, **1953**, 2939.
19. J.T. van Germert and P.R. Wilkinson, *J. Phys. Chem.*, **68**, 645 (1964).
20. A.P. Stepovik, A.K. Shilova and A.E. Shilov, *Dokl. Akad. Nauk SSSR*, **148**, 122 (1963).
21. F.C. Diyakovskii and A.E. Shilov, *Zh. Fiz. Khim.*, **41**, 2512 (1967).
22. R. Cramer, *J. Amer. Chem. Soc.*, **87**, 4717 (1965).
23. A.D. Ketley, L.P. Fisher, A.T. Berlin, C.R. Morgan, E.H. Gorman and T.R. Steadman, *Inorg. Chem.*, **6**, 657 (1967).
24. A. Yamamoto, T. Shimizu and S. Ikeda, *Makromol. Chem.*, 1971 (in press); T. Yamamoto, A. Yamamoto and S. Ikeda, presented at 23rd Annual Meeting of Japan Chem. Soc., 1969; *J. Amer. Chem. Soc.*, **93**, no. 6 (in press); Y. Yamamoto, *Dr. Eng. Thesis, Tokyo Inst. Technology*, 1971.
25. J. Brandrup and E.H. Imergut, *Polymer Handbook*, p. 341, Interscience, 1966.
26. M.J.S. Dewar, *Bull. Soc. France*, **18**, C79 (1951).
27. A. Streitwiesser Jr., *Molecular Orbital Theory for Organic Chemists*, Wiley, 1961.
28. T. Kagiya, Y. Sumida and T. Nakata, *Bull. Chem. Soc. Japan*, **41**, 2239 (1968).

Chapter 8

Electron spin resonance spectra of Ziegler-Natta catalyst systems

Electron spin resonance absorption provides a powerful tool for investigating the structures of complexes involving transition metal atoms, and since the work of Shilov[1)] many Ziegler-Natta systems have been studied by this method. The results of some of these studies are discussed below.

8.1 HETEROGENEOUS ZIEGLER-NATTA SYSTEMS

8.1.1 ESR spectra of $TiCl_3$

Ono and Keii[2)] found two absorption spectra from $TiCl_3$ (AA-grade, 21 m^2/g); one is sharp (ΔH_{msl} = 15 gauss) with a g value of 1.94 and the other is very broad. The number of unpaired electrons in the former was found to be $2 \times 10^{16}/g$. Aithough Takegami *et al.*[3)] questioned the existence of this absorption in HA-grade $TiCl_3$ (hydrogen reduced), Okura *et al.*[4)] confirmed the presence of the absorption ($g = 1.94$) in milled samples of both types of $TiCl_3$. The latter authors examined several kinds of commercial $TiCl_3$, including A- and H-grades of Stauffer Chemical Co., and TAC-grade of the Toho Titanium Co. Samples of A-grade (aluminium reduced, without activation) and H-grade (hydrogen reduced, without activation) showed no spectrum, but after ball-milling for 100 hr (porcelain or stainless steel ball mill), the samples gave the same spectrum ($g = 1.94$) as obtained from TAC-grade $TiCl_3$ (aluminium reduced, activated, with surface area about 10 m^2/g). The spectra are shown in Fig. 8.1.

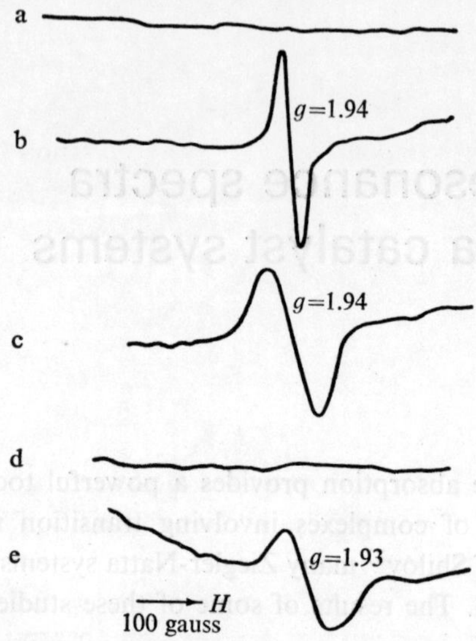

Fig. 8.1. ESR spectrum of $TiCl_3$ after grinding. (a, Stauffer A-grade (unground); b, ground A-$TiCl_3$; c, TAC-$TiCl_3$; d, H-$TiCl_3$ (unground); e, H-$TiCl_3$.)

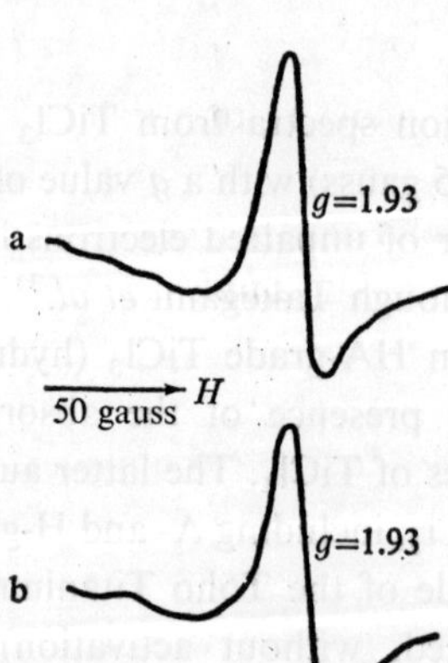

Fig. 8.2. ESR spectrum of A-$TiCl_3$ after addition of olefin. (a, ethylene; b, propylene.)

The spectrum with $g = 1.94$ can be regarded as characteristic of activated $TiCl_3$. It was found that the intensity was strongly increased by the addition of small amounts of $AlCl_3$,[5] and it was also observed that A- or H-grade $TiCl_3$ (not activated) give the same spectrum in the presence of ethylene, propylene, or alkylaluminiums such as $AlEt_3$, $AlEt_2Cl$, and $AlEtCl_2$ (Fig. 8.2). This absorption could not be attributed to any surface complex, but appears to be due to Ti (III) formed on the solid surface by grinding, or by reduction with olefin or alkylaluminium. Okura *et al.*[4] suggested a chlorine deficient structure for Ti (III).

8.1.2 ESR spectra of $TiCl_3$/alkylaluminium

Adema *et al.* examined the systems $TiCl_4/AlEt_2Cl$[6] and $TiCl_4/AlEtCl_2$[7] and concluded that there was no spectrum from the solid phase, but two absorptions by the liquid phase. Ono and Keii[2] observed similar spectra with $TiCl_4/AlEt_3$, though this changed with increasing contact time as shown in Fig. 8.3. However, on separation of the two phases by centrifugation, the latter authors observed an absorption by the solid phase (Fig. 8.4). This was found to be identical to that observed for the solid phase from AA-grade $TiCl_3/AlEt_3$ (Fig. 8.5*b*). Fig. 8.5*c* shows the spectrum due to the liquid phase of the $TiCl_3/AlEt_3$ system, for which $g = 1.972$. This has been attributed to a soluble complex of $AlEt_3$ with $TiCl_4$ occluded in the bulk of the $TiCl_3$, but more systematic studies by Okura *et al.*[4] indicate that this is not so. Fig. 8.6 shows the ESR spectra of AA-grade $TiCl_3$ (specific surface area 10 m^2/g) with (a) $AlEt_3$, (b) $AlEt_2Cl$ and (c) $AlEtCl_2$ under nitrogen (1) and propylene (2), respectively. All samples of $TiCl_3$ were evacuated at 200 °C to remove any traces of $TiCl_4$. It can be seen that the intensities of the absorptions with $g = 1.97$ increased considerably in the presence of olefins, and were proportional to the polymerization activity of the catalyst system, i.e.

$$AlEt_3 > AlEt_2Cl > AlEtCl_2 \simeq 0 \tag{8.1}$$

The intensity of the absorption with $g = 1.94$ remained unchanged with contact time, irrespective of the presence or absence of propylene, as shown in Fig. 8.7. The number of spins for the system TAC-grade $TiCl_3/AlEt_3$ was found to be 2×10^{19}/g-$TiCl_3$, corresponding to $2 \times 10^{14}/cm^2$.

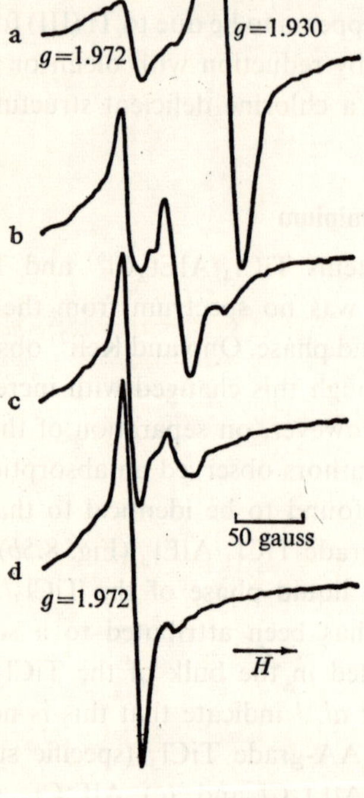

Fig. 8.3. Changes in ESR spectrum of AA-$TiCl_3$ (21 m^2/g)/$AlEt_3$ with contact time from *a* (5 min) to *d* (*ca.* 1 hr.)

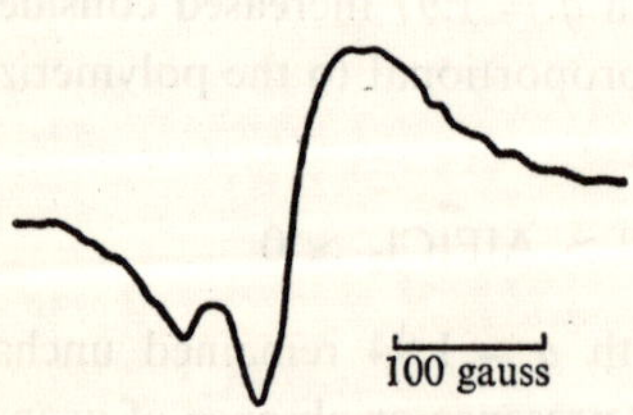

Fig. 8.4. ESR spectrum of the precipitate from AA-$TiCl_3$/$AlEt_3$ mixture.

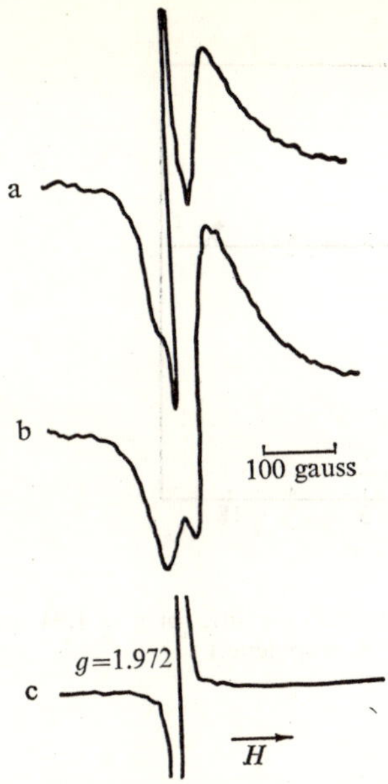

Fig. 8.5. ESR spectrum of $TiCl_3/AlEt_3$ system. (a, AA-$TiCl_3/AlEt_3$; b, solid phase; c, liquid phase.)

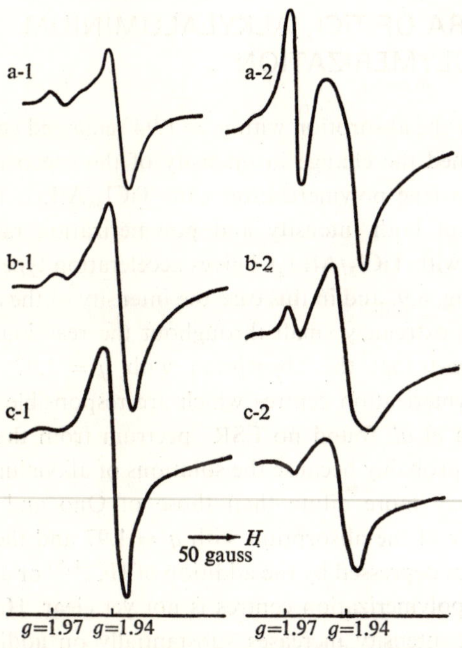

Fig. 8.6. ESR spectrum of AA-$TiCl_3$ after addition of alkylaluminium and propylene. (1, before addition of propylene; 2, after addition of propylene; a, $AlEt_3$, b, $AlEt_2Cl$; c, $AlEtCl_2$.)

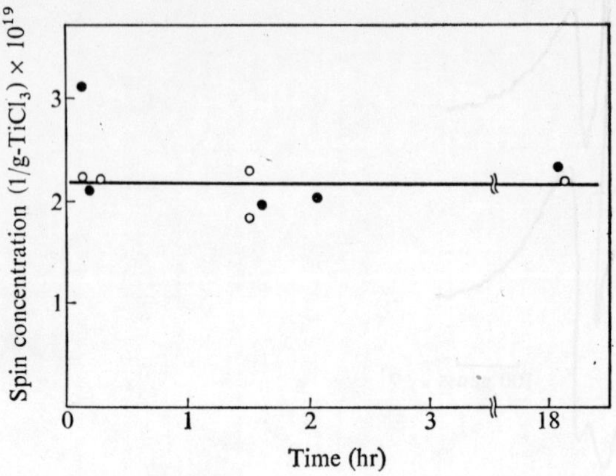

Fig. 8.7. Dependence of spin concentration on time for g = 1.94 spectrum. (in the presence (○) and absence (●) of propylene.)

8.2 ESR SPECTRA OF $TiCl_3$/ALKYLALUMINIUM DURING POLYMERIZATION

By assuming that the absorption with $g = 1.94$ remained constant, Okura *et al.*[4] examined the change in intensity of the spectrum with $g = 1.97$ during propylene polymerization with $TiCl_3/AlEt_3$. Fig. 8.8 shows the variation of both intensity and polymerization rate with time. Polymerization with $TiCl_3/AlEt_2Cl$ gives acceleration type kinetic curves, as shown in Fig. 8.9, and in this case the intensity of the absorption with $g = 1.97$ is extremely small throughout the reaction. These results strongly suggest that the absorption with $g = 1.97$ can be attributed to the polymerization centres which are responsible for the initial activity. Okura *et al.* found no ESR spectrum from the liquid phase of the catalyst, probably because the solutions of alkylaluminium used were considerably more dilute than those of Ono and Keii.[2]

Both the intensity of the absorption with $g = 1.97$ and the initial polymerization rate are depressed by the addition of HCl[4,8] or amine.[9] The structure of the polymerization centres is not yet clear. However, the observations that intensity increases substantially on addition of ethylene or propylene, and that the shape of the spectrum remains unchanged irrespective of the nature of the olefin suggest that olefin

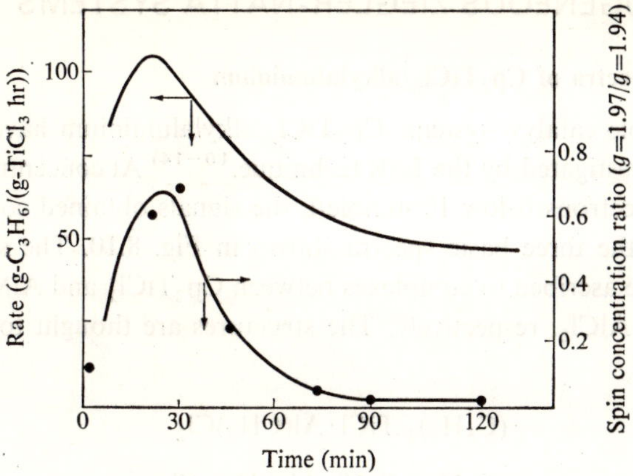

Fig. 8.8. Dependence of propylene polymerization rate ($TiCl_3/AlEt_3$) and spin concentration ratio (g = 1.97/g = 1.94) on time. ([TAC-$TiCl_3$] = [$AlEt_3$] = 3.3 mmole/250 ml *n*-heptane at 40°C. P_{Pr} = 70 cmHg. Polymerization rate (——) and spin concentration ratio (—●—).)

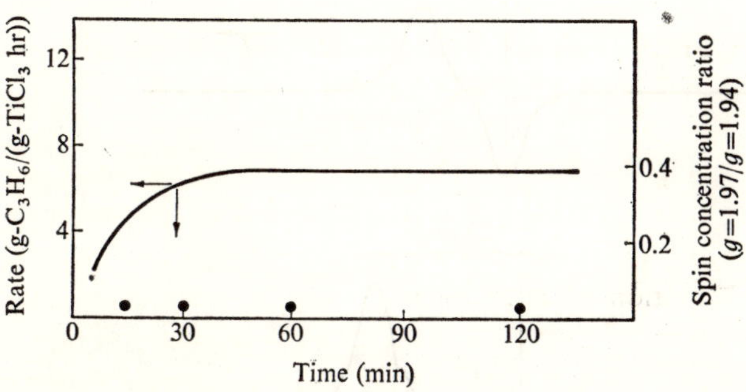

Fig. 8.9. Dependence of spin concentration (g = 1.97, based on g = 1.94 as constant) and rate of polymerization on time for $TiCl_3/AlEt_2Cl$. (Conditions as for Fig. 8.8.)

monomer units participate in the formation of polymerization centres. This is compatible with the kinetic conclusions. When alkylaluminium is added to $TiCl_3$ under nitrogen, a spectrum of slight intensity appears, but this can be attributed to the formation of some ethylene by reaction of $TiCl_3$ and alkylaluminium.

The structures of soluble titanium complexes formed in homogeneous Ziegler-Natta systems may throw some light on the above problems, and will be discussed in the next section.

8.3 HOMOGENEOUS ZIEGLER-NATTA SYSTEMS

8.3.1 ESR spectra of Cp_2TiCl_2/alkylaluminium

The soluble catalyst systems Cp_2TiCl_2/alkylaluminium have been extensively investigated by the ESR technique.[10–14] At concentrations of unpaired electrons below 10 mmole/l, the signals obtained could be resolved into the three basic spectra shown in Fig. 8.10. The spectra a, b and c were ascribed to complexes between Cp_2TiCl_2 and $AlMe_2Cl$, $AlMeCl_2$ and $AlCl_3$, respectively. The structures are thought to be as follows.

$$\begin{array}{ll} \text{a} & (C_5H_5)_2\ TiCl\cdot Al(CH_3)Cl_2 \\ \text{b} & (C_5H_5)_2\ TiCl\cdot Al(CH_3)_2Cl \\ \text{c} & (C_5H_5)_2\ TiCl\cdot AlCl_3 \end{array} \tag{8.2}$$

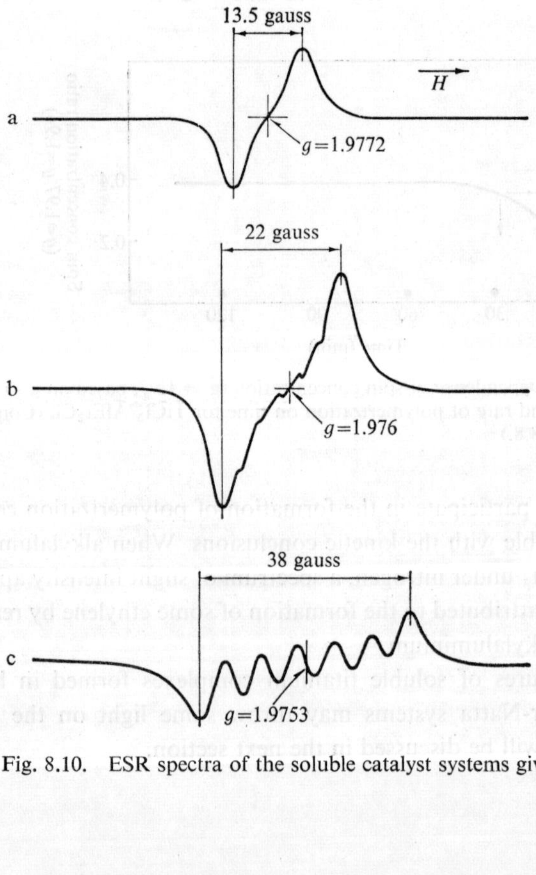

Fig. 8.10. ESR spectra of the soluble catalyst systems given in Eq. 8.2.

8.3.2 ESR spectra of $Ti(OBu)_4$/alkylaluminium

Dzhabiev *et al.*[15] examined the $Ti(OBu)_4/AlEt_3$ system ([A]/[Ti] = 2) and obtained a spectrum with $g = 1.957$ and having a hyperfine structure of eleven components, as shown in Fig. 8.11. The structure

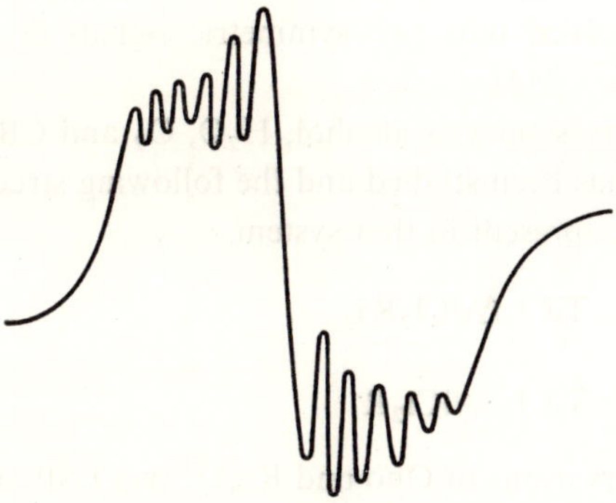

Fig. 8.11. ESR spectrum obtained from a mixture of $Ti(OBu)_4$ and $AlEt_3$.[15]

given in Eq. 8.3 was proposed for the complex

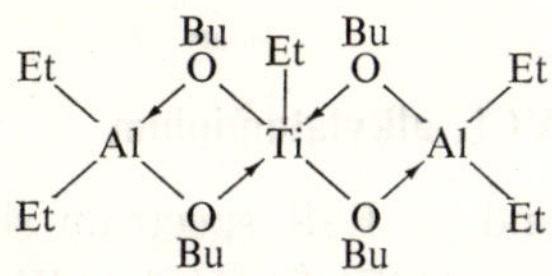

(8.3)

A similar structure has also been put forward by Angeliscu *et al.*[16] who examined the $Ti(OBu)_4/TiEt_2Cl$ system.

Takeda *et al.*[17] observed two spectra, with $g = 1.978$ and 1.962, from $Ti(OBu)_4/AlEt_3$ in the presence of styrene. As styrene polymerization progressed, the intensity of the spectrum with $g = 1.978$ diminished, whereas the intensity of the other increased. It was concluded that the active species in the system was $(BuO)_3TiEt$.

8.3.3 ESR spectra of $TiCl_3$/alkylaluminium

The ESR spectra of soluble complexes formed by $TiCl_4$ and $AlEt_2Cl$ or $AlEtCl_2$ have been studied extensively by Adema *et al.*[6,7] In the case of $TiCl_4/AlEtCl_2$, two kinds of ESR active species, S_1 and S_2 were found, where $[S_1] > [S_2]$. The rate of formation of S_1 could be expressed as

$$\frac{d[S_1]}{dt} = k[\mathrm{Ti}]_0^{1/2}[\mathrm{Al}]_0^{3/2} \tag{8.4}$$

$$k = k_0 e^{-12.8\mathrm{kcal}/RT} \tag{8.4a}$$

The ESR signals observed from $TiCl_4/AlEtCl_2$ consist of a symmetrical signal ($\Delta H_{msl} = 22$ gauss, $g = 1.925$) and an asymmetric signal which could be resolved into two symmetric signals (S_1 with $g = 1.925$ and S_2 with $g = 1.933$).

The influence of additives such as alcohol, H_2O, O_2 and CBr_4 on the $TiCl_4/AlEt_2Cl$ system has been studied and the following structures proposed for the complexes present in that system.

$$\begin{array}{ll} S_1 & TiCl_3(AlCl_2R)_2 \\ S_2 & TiCl_3 \cdot AlCl_2R \end{array} \tag{8.5}$$

According to the observations of Ono and Keii,[2] two ESR signals are obtained from the liquid phase of the $TiCl_4/AlEt_3$ system; one has $\Delta H_{msl} = 19$ gauss, $g = 1.924$ and the other has $\Delta H_{msl} = 19$ gauss, $g = 1.972$. The former appears to be similar to that found by Adema *et al.*

8.3.4 ESR spectra of VCl_4/alkylaluminium

Natta *et al.*[18] found an ESR spectrum similar to that of VCl_2 from the systems $V(acac)_3/AlEt_2Cl$, $VCl_4/AlEt_2Cl$/anisole and $VCl_4/AlEt_3$ at room temperature. Angeliscu *et al.*[19] observed a spectrum with eight components from a $VO(acac)_2/AlEt_2Cl$ system, and Shilov *et al.*[20] also found an eight component spectrum with $g = 1.992$ from a $VCl_4/AlEt_3$/benzene system. Substitution of C_2D_5 for C_2H_5 caused no change in the spectrum, and on this basis, Shilov *et al.* supposed that the spectrum arose from the following complex species.

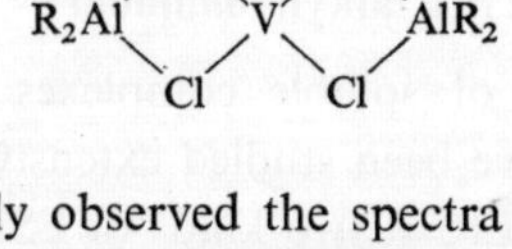

Okura[21] separately observed the spectra of the solid and liquid phases separated from VCl_4 and $AlEt_3$ (A), $AlEt_2Cl$ (B) or $AlEtCl_2$ (C). The resulting spectra are shown in Fig. 8.12. The liquid phase from

$VCl_4/AlEt_3$ shows an eight component spectrum ($\Delta H_{msl} = 44$ gauss, $g = 1.99$) due to the 7/2 nuclear spin of V[15], and the solid phase gives

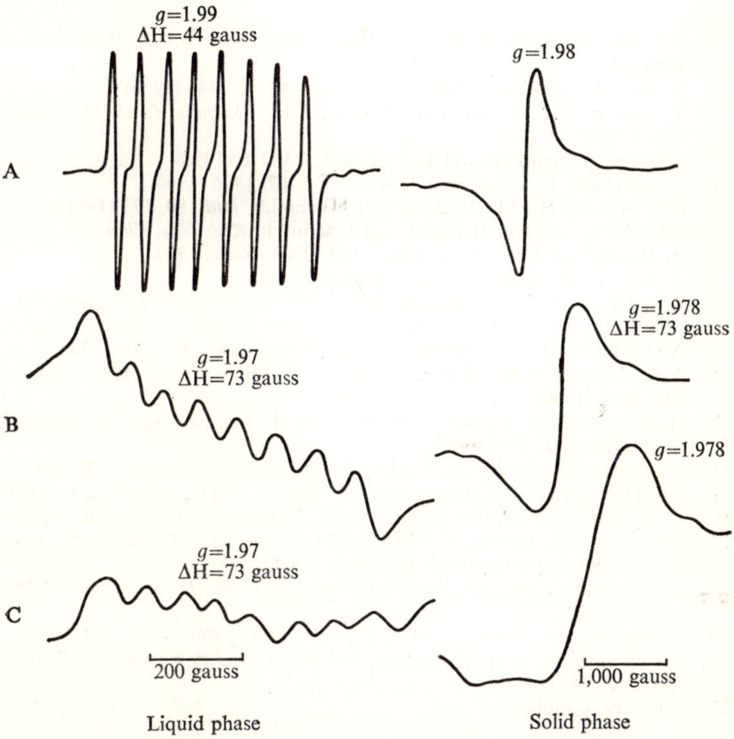

Fig. 8.12. ESR spectra of the liquid and solid phases from VCl_4 and $AlEt_3$ (A), $AlEt_2Cl$ (B) and $AlEtCl_2$ (C).

a spectrum with $g = 1.98$. From the other mixtures, an eight component spectrum with $\Delta H = 73$ gauss, $g = 1.97$ was obtained from the liquid phase and a spectrum with $g = 1.978$ from the solid phase. Since the spectral width, ΔH, corresponds to the electron density around V, these results suggest that the reducing powers of the alkylaluminiums are in the order

$$AlEt_3 < AlEt_2Cl \simeq AlEtCl_2$$

This is, however, the reverse of the order accepted by many authors.

REFERENCES

1. A.E. Shilov and N.N. Bubnov, *Izvest. Akad. Nauk SSSR*, *Otdelkhim*, *Nauk*, **1958**, 81.
2. Y. Ono and T. Keii, *J. Polymer Sci.*, part A1, **4**, 2441 (1966).
3. Y. Takegami, T. Suzuki and M. Fujii, *Kogyo Kagaku Zasshi* (Japanese), **71**, 1554 (1968).
4. I. Okura, Y. Sendoda and T. Keii, *ibid.*, **73**, 276 (1970).
5. S. Kanetaka, T. Takagi and T. Keii, *ibid.*, **67**, 1436 (1964).
6. E.H. Adema, H.J.M. Bartelink and J. Schmidt, *ibid.*, **80**, 173 (1961).
7. E.H. Adema, H.J.M. Bartelink and J. Schmidt, *Rec. Trav. Chim.*, **81**, 73 (1962).
8. Y. Hattori, *M. Eng. Thesis*, *Tokyo Inst. Technology*, 1969.
9. T. Keii, Y. Doi and I. Okura, unpublished paper.
10. A.E. Shilov, A.K. Zefirova and N.N. Tikhomirova, *Zhr. Fiz. Khim.*, **33**, 2113 (1959).
11. A.H. Maki and E.W. Randall, *J. Amer. Chem. Soc.*, **82**, 4109 (1960).
12. A.K. Zefirova, N.N. Tikhomirova and A.E. Shilov, *Dokl. Akad. Nauk SSSR*, **132**, 1082 (1960).
13. H.J.M. Bartelink, H. Bos, J. Schmidt, C.H. Vrissen and E.H. Adema, *Rec. Trav. Chim.*, **81**, 225 (1962).
14. P.E. Allen, J.K. Brown and R.M.S. Obaid, *Trans. Faraday Soc.*, **85**, 1808 (1963).
15. T.S. Dzhabiev, R.D. Sabirova and A.E. Shilov, *Kinetika i Kataliz*, **5**, 441 (1964).
16. E. Angeliscu, C. Nicolau and Z. Simon, *J. Amer. Chem. Soc.*, **88**, 3910 (1966).
17. M. Takeda, K. Iimura, Y. Nozawa, M. Kuru, and M. Watanabe, presented at 13th Symposium on High Polymers, 1964.
18. A. Zambelli, G. Natta, I. Pasquon and R. Signorini, *J. Polymer Sci.*, **C16**, 2485 (1967).
19. E. Angeliscu, C. Nicolau and Z. Simon, *J. Amer. Chem. Soc.*, **88**, 3910 (1966).
20. S.V. Shulyndin, N.N. Tikharova, A.E. Shilov and A.K. Shilova, *Zhur. Struk. Khim.*, **2**, 740 (1961).
21. I. Okura, unpublished work.

Chapter 9

Mechanism of polymerization by Ziegler-Natta catalysts

In the previous chapters, the polymerization of ethylene and propylene by heterogeneous Ziegler-Natta catalysts has been discussed in terms of kinetic behavior and the corresponding kinetic models. The detailed mechanism of polymerization will now be considered in the light of these results.

9.1 THE ROLE OF THE $TiCl_3$ SURFACE

It is known that isotactic polypropylene is formed only with heterogeneous Ziegler-Natta catalysts. The surface of the catalyst can therefore be regarded not only as the source of the Ti complexes which act as polymerization centres, but also as the substrate upon which isotactic polymerization can take place.

The first attempt to account for isotactic polymerization is due to Friedlander and Oita,[1] who proposed the "bound-radical" chain growth mechanism shown in Fig. 9.1. A more comprehensive mechanism, in which the role of the catalyst surface is taken into account, is that of Furukawa and Tsuruta[2] (Fig. 9.2). Both the above mechanisms assume that the catalyst surface is a cleavage or net plane of the $TiCl_3$ crystal, whereas Natta[3] considered that the adsorption centres for alkylaluminium are located on edge structures in the $TiCl_3$ crystal (Fig. 9.3). These two suggestions have been investigated by Rodriguez and Gabant,[4] who took electron micrographs of a $TiCl_3$ catalyst surface after polymerization had taken place for a short time. The surface of the (1000) plane is shown in Fig. 9.4. It can be seen that many dots are

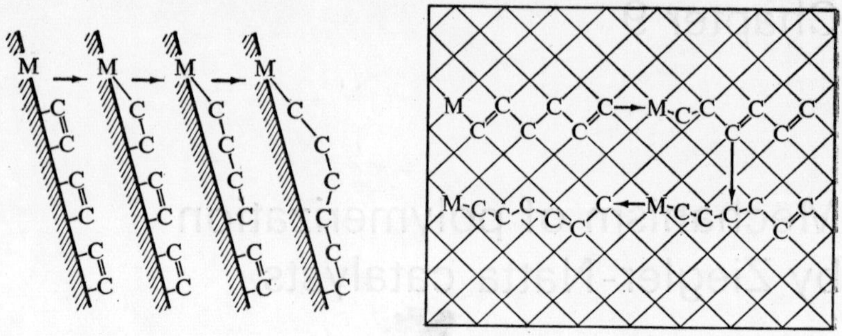

(a) (b)

Fig. 9.1. "Bound radical" polymerization mechanism proposed by Friedlander and Oita.[1] (a, cross section; b, surface view.)

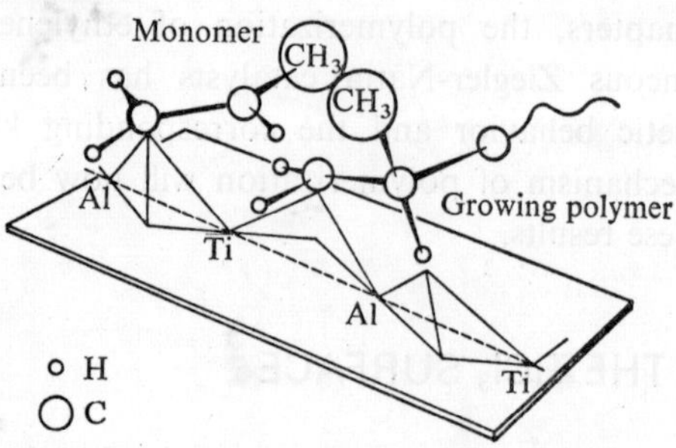

Fig. 9.2. Mechanism of Furukawa and Tsuruta.[2]

located in a spiral on the plane, and it was concluded by Rodriguez and Gabant that only the exposed Ti atoms were responsible for the formation of polymerization centres, rather than all of the surface atoms.

It is well known that crystal growth generally takes place along a spiral on a plane perpendicular to the direction of growth, and that the adsorption of molecules onto that plane is the rate determining step.[5] It seems possible that α-olefin polymerization may occur via a similar mechanism, but further work on polymerization by single crystals of $TiCl_3$ would be necessary to confirm this. The α-$TiCl_3$ crystals used by Rodriguez and Gabant were prepared by the reduction of $TiCl_4$ with hydrogen.

Okura and Keii[6] have observed the surface of a large polycrystalline sample of α-$TiCl_3$ during the gas-phase polymerization of propylene with $AlEt_2Cl$ at room temperature. Fig. 9.5 shows an electron micro-

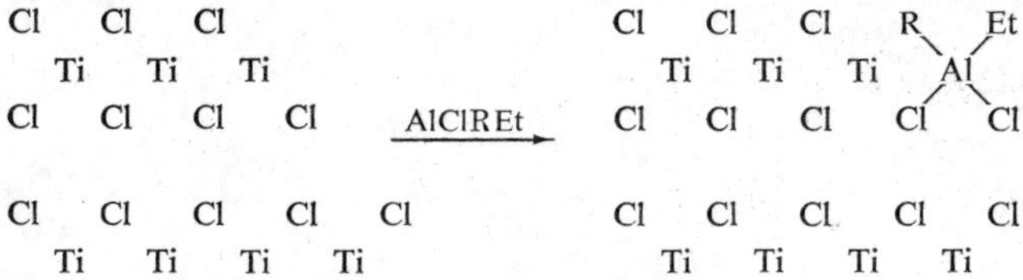

Fig. 9.3. Adsorption of AlCl(R)Et on $TiCl_3$ crystal.[3]

Fig. 9.4. Surface view of (1000) plane of α-$TiCl_3$ after polymerization with $AlMe_3$ for a short time.[4]

graph obtained after 15 min polymerization. Several types of structure (B, large colony; C, linear array) can be seen in addition to the spiral structure (A) on the surface of the sample, where many crystal planes are exposed.

Samples of $TiCl_3$ prepared by the reduction of $TiCl_4$ with alkylaluminium have been examined microscopically by Hock,[7] who found that the particles (~ 30μ) of $TiCl_3$ were composed of fine polygonal leaflets having a thickness of 300–1000 Å. Propylene polymerization was found to occur on the surface of these leaflets.

The $TiCl_3$ used in Ziegler-Natta polymerization is usually activated by prolonged grinding. In addition to the change from α- or γ-$TiCl_3$ to the δ-modification, grinding markedly disrupts the crystal bonding

Fig. 9.5. Electron micrograph of the $TiCl_3$ surface during gas-phase polymerization. (H-$TiCl_3$/$AlEt_2Cl$ after 10 min polymerization at room temperature.)

structure, at least on the surface of the crystals, as indicated by the appearance of an esr signal ($g = 1.94$) due to Ti (III) after grinding (or on contact with olefin or alkylaluminium).[6,7] The surface of ground $TiCl_3$ may therefore be rather rough, so that the number of active sites (e.g. exposed Ti) can be considered to be of the same order of magnitude as the number of surface Ti atoms. This is in agreement with the kinetic conclusions (4.3.1).

Okura *et al.*[8] examined a polymerizing system microscopically, and found that both fibrous and lamellar polymer growths could be observed (Fig. 7.11). It has often been supposed that two types of polymerization centres exist in $TiCl_3$, and in addition to these direct observations, Boor[9] assumed the existence of two different structures for active centres to account for the formation of both isotactic and atactic polymers during Ziegler-Natta polymerization. Many other authors have also supported the existence of catalyst surface heterogeneity.

Arlman[10] considered that the results obtained by Rodriguez and Gabant (Fig. 9.4) indicated that the active sites are localized on the $(10\bar{1}0)$ surface of the α-$TiCl_3$ crystal. He suggested that these active sites were surface chlorine vacancy sites, which should be formed to keep the crystal in an overall electroneutral state. Such vacancies have been assumed to be the active sites in Cossee's mechanism, which will be discussed later.

9.2 EARLY WORK ON THE POLYMERIZATION MECHANISM

Several of the mechanisms proposed during the earlier stages of research centred around free radical propagation. Friedlander and Oita[1] considered that adsorbed organometallic compounds or titanium alkyls functioned as polymerization initiators (M in Fig. 9.1). The adsorption of organometals and formation of the titanium alkyls were assumed to occur as follows.

$$\begin{aligned} TiCl_3 + nMR &\rightarrow TiCl_3(MR)_n \\ TiCl_3 + MR &\rightarrow TiCl_2R + MCl \\ TiCl_3 + 2MR &\rightarrow TiClR_2 + 2MCl \\ TiCl_2 + 2MR &\rightarrow TiR_2 + 2MCl \end{aligned} \tag{9.1}$$

The growing polymer chains were assumed to be "bound radicals", as in Fig. 9.1, or "bound ion-radical" in Friedlander's "replica polymerization" mechanism.

Several authors[11-15] considered the formation of titanium alkyls by organometallic compounds. According to Nenitzescu *et al.*,[11] titanium alkyls are formed by reactions of the following type

$$\begin{aligned} TiCl_4 + RNa &\rightarrow NaCl + RTiCl_3 \\ RTiCl_3 + RNa &\rightarrow NaCl + R_2TiCl_2 \\ R_2TiCl_2 + RNa &\rightarrow NaCl + R_3TiCl \end{aligned} \tag{9.2}$$

and then yield radicals which induce the polymerization. Ludlum *et al.*[12] considered the Ti (II) state to be the most active, and proposed the mechanism given in Eq. 9.3

$$\begin{array}{c} CH_2{=}CH_2 \\ \downarrow \\ Cl\ Ti(C_2H_4)_nR \end{array} \longrightarrow Cl\ Ti(C_2H_4)_{n+1}R \tag{9.3}$$

McGowan *et al.*[13] preferred an ion-radical mechanism

$$R{-}\overset{|}{\underset{|}{Ti}}+CH_2{=}CH_2 \rightleftharpoons [R{-}\overset{|}{\underset{|}{Ti}}{}^{\oplus}][^{\ominus}CH_2{-}\dot{C}H_2] \qquad (9.4)$$

and Topchev *et al.*[14,15] proposed the following mechanism

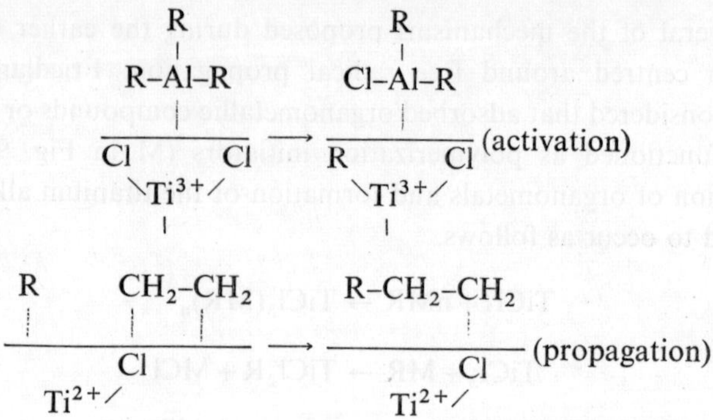

(9.5)

Natta[3] favoured a coordinated anionic polymerization mechanism, based on the observations that the products of monomer reactivity ratios for ethylene and propylene $[r(C_2H_4) \times r(C_3H_6)]$ are close to unity, and that $r(C_2H_4) \gg r(C_3H_6)$, as shown in Table 9.1. On the

TABLE 9.1 Copolymerizations of ethylene and propylene[3]

Catalyst	$r(C_2H_4)$	$r(C_3H_6)$	$r(C_2H_4) \times r(C_3H_6)$
$TiCl_4 + AlEt_3$	37.36 ± 1.31	0.032 ± 0.001	1.20
$VOCl_3 + AlEt_3$	17.95 ± 0.27	0.065 ± 0.002	1.15
$VCl_4 + AlEt_3$	7.08 ± 0.20	0.088 ± 0.003	0.63

other hand, the cationic nature of Ziegler-Natta catalysts is indicated by the observation that they produce stereospecific polymerization of isobutyl vinyl ether at $-78\,°C$.[16,17] As described in 7.1, the acidity of $TiCl_3$/alkylaluminium systems depends on the nature of the alkylaluminium and its molar ratio to $TiCl_3$. Furthermore, the charge type of the catalyst is also dependent on the nature of the monomer molecule.

Furukawa and Tsuruta[2] proposed a polymerization centre involving two metal atoms, Ti (electron acceptor) and Al (donor), as shown in Eq. 9.6. This is based on redox-redoxy (acid-base) catalyst theory.

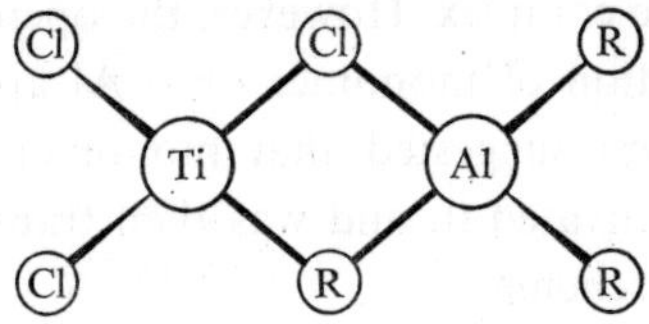

(9.6)

9.3 THE BIMETALLIC MECHANISM

A bimetallic polymerization mechanism (i.e. active centre of the type shown in Fig. 9.6) was first proposed by Natta,[3] and has since received considerable support. The observation that the stereospecificity of the catalyst system depends markedly on the nature of the cocatalyst (4.1.2) provides strong evidence for a mechanism of this type. From a kinetic point of view, support for this approach is provided by the variation in activation energy for propylene polymerization with different cocatalysts (e.g. 10 kcal/mole[18] for $TiCl_3/AlEt_3$ and 2.7 kcal/mole[19] or 4.4 kcal/mole[20] for $TiCl_3/ZnEt_2$), and by the difference in the types of kinetic curve obtained with different cocatalysts [e.g. $AlEt_3$ and $AlEt_2Cl$, as described in 2.1].

Natta *et al.*[21,22] studied the soluble bimetallic complex shown in Eq. 9.7, which catalyses ethylene polymerization.

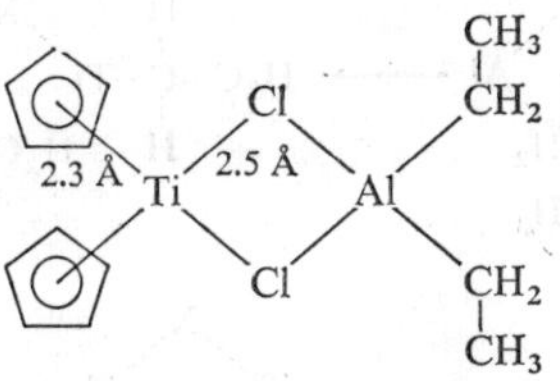

(9.7)

They found that polymers produced with this complex had the end groups shown in Table 9.2, and concluded that polymerization involved

TABLE 9.2 End group of polyethylene produced with $Cp_2TiR_2'-AlR_3$

Catalyst	End group
$Cp_2TiCl_2 + Al(C_6H_5)_3$	C_6H_5-
$Cp_2Ti(C_6H_5)_2 + AlEt_3$	Et-
$Cp_2Ti(C_6H_5)_2 + Al(C_6H_5)_3$	C_6H_5-

the Al–R bond of the complex. However, the opening of an Al–C bond formed by coordination of monomer onto Al appears to be difficult, and Natta[3] therefore suggested that monomer initially coordinated onto the Ti atom, activated it, and was then transferred to Al according to the following scheme

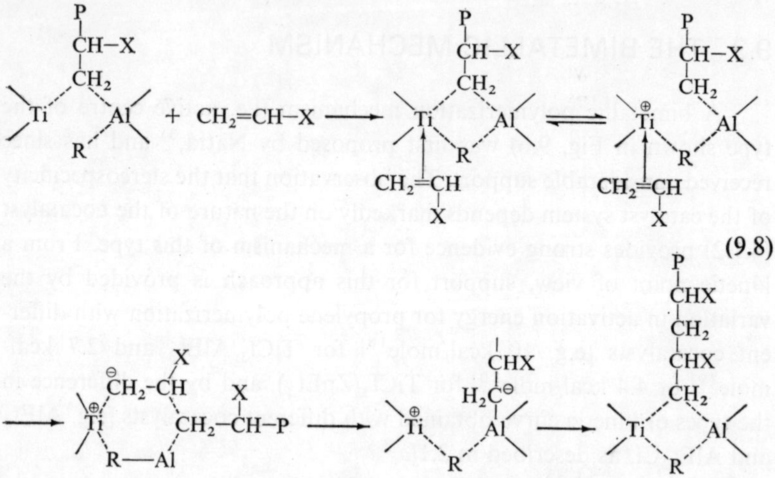

(9.8)

Essentially the same mechanism has been proposed by Patat and Sinn[23]

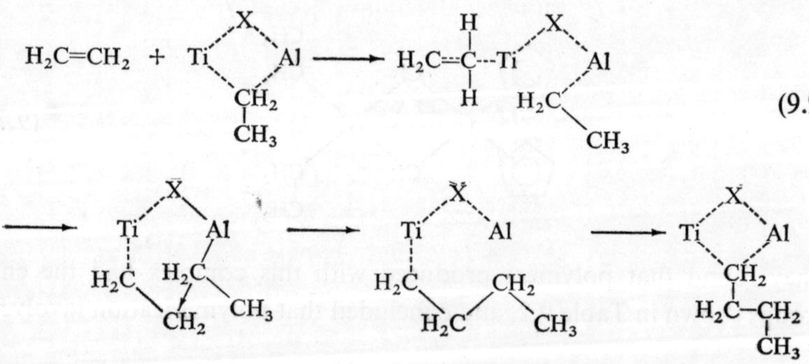

(9.9)

and Uelzmann[24] has also proposed a bimetallic mechanism.

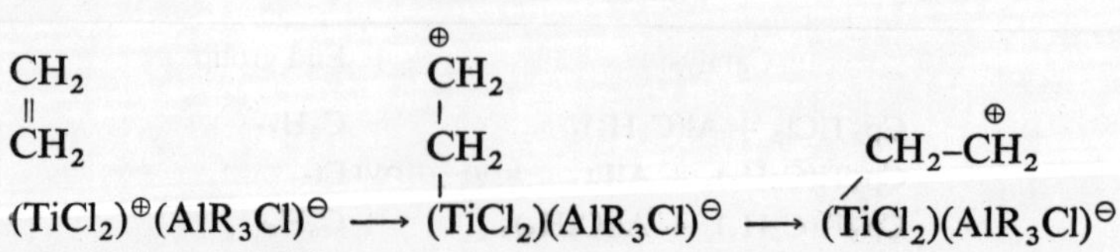

$$\longrightarrow (TiCl_2)^{\oplus}(\overset{CH_2CH_2R}{\diagup}AlR_2Cl)^{\ominus} \xrightarrow{nC_2H_4} (TiCl_2)^{\oplus}(\overset{(CH_2CH_2)_{n+1}R}{\diagup}AlR_2Cl)^{\ominus} \quad (9.10)$$

Other authors to favour a bimetallic mechanism include Huggins[25] and Boor.[26]

The complex $Cp_2TiR'_2$ has been investigated further by Karapinka and Carrick,[27] who found that interchange of R′ and R between the complex and AlR_2Cl cccurs rapidly. This result brings into question the validity of Natta's end group analysis studies. Recently Olive[28,29] has investigated the complex shown in Eq. 9.11.

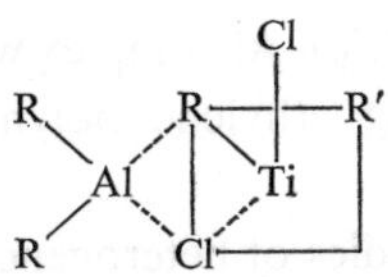

(9.11)

It was found that when R = Me and R′ = Bu, no interchange took place, but with R = Bu and R′ = Me, interchange was observed. Furthermore, it was confirmed that reaction of ethylene with alkyl groups coordinated to Ti did occur. These results further reduce the apparent validity of the bimetallic mechanisms previously illustrated, but the fact remains that the bimetallic mechanism provides a simple explanation for the isotacticity of Ziegler-Natta polymers.

9.4 THE MONOMETALLIC MECHANISM

Monometallic polymerization mechanisms assume that Ti–R acts as the polymerization centre, and that the role of alkylaluminium is to alkylate $TiCl_3$. An increasing amount of evidence has been found to support this kind of mechanism.

It has been confirmed by many authors that transition metals can be alkylated by organometals. Beerman and Bestian[30] detected the formation of $MeTiCl_3$ or Me_2TiCl_2 during the reaction of $TiCl_4$ with $AlMe_3$ or $AlMe_2Cl$, and Arlman[31] found that $ScCl_3$ could be alkylated by $ZnEt_2$. Similarly, the formation of an M–R bond during the reaction of $M(acac)_2$ with $AlR_2(OEt)$ and dipyridyl (M = Fe, Co, Ni) has been reported by Yamamoto *et al.*,[32,33] who isolated the product,

$MR_2(dipy)_2$, and identified the M–R bond by NMR spectroscopy.

Further evidence for a monometallic mechanism has arisen from kinetic research into soluble complexes. Long and Breslow[34)] found a complex, $Cp_2TiClEt \cdot EtAlEt_2$ with a Ti–Et bond in a mixture of Cp_2TiCl_2 and $AlEt_2Cl$, and were able to correlate the concentration of this complex with the initial rate of ethylene polymerization by the mixture. The homogeneous system $Cp_2TiMeCl/AlMeCl_2$ has been extensively studied by Shilov and coworkers,[35)] who obtained the most striking result that during the electrolysis of $Cp_2TiCl_2/AlMe_2Cl$ in dichloroethane, ethylene polymerization was found to occur at the cathode, where Cp_2TiMe^+ was present, but not at the anode, where $AlMeCl_2^-$ was present. On the other hand, Reichert and Schubert[36)] inferred the existence of a Ti(IV)–Al complex with a bridged ethyl group as the active species during ethylene polymerization by $Cp_2TiCl_2/AlEt_2Cl$ in benzene.

A number of kinetic studies of heterogeneous catalyst systems also indicate that a monometallic mechanism is preferable. Karol and Carrick[37)] copolymerized ethylene and propylene using catalysts based on transition metal compounds and $Al(i\text{-}Bu)_3$, $Zn(C_6H_5)_2$, $Zn(n\text{-}Bu)_2$ or $MeTiCl_3$, and found that the monomer reactivity ratios depended only on the nature of the transition metal compound. The reactivity of propylene increased with increasing transition metal electronegativity, in the order

$$VCl_4 > VOCl_3 > TiCl_4 > ZrCl_4 > HfCl_4$$

Analogous results were obtained by Danusso and Sianesi[38)] for the copolymerization of styrene and *p*-methylstyrene at 60 °C.

The above results obviously indicate that the transition metal centre participates in the chain growth mechanism, but Schindler[39)] suggested that the cocatalyst metal atom must also be involved, on the basis of the observation that transfer and hydrogen exchange reactions in the presence of deuterium were affected by the nature of the cocatalyst. These results were confirmed by Keii and Kojima[40)] (see 7.3), who ascribed them to the differences in the reducing powers of the cocatalysts for $TiCl_3$.

The most significant evidence for a monometallic mechanism is the observation that isotactic polymerization of propylene was found to occur with $TiCl_3$/tri-*n*-butyl amine in the absence of any organometallic compounds, as reported by Boor.[41] The fact that the polymerization activity was so low that it could only be measured after 28 days was presumed to be due to the small amount of Ti–R present. Matlack and Breslow[42] have also shown that many transition metals (V, Nb, Ta, La, Le, Th, U), mixtures of C_2H_5Br with Ti, V or Th, hydrides of Ti or Zr, and H_2 with Ti or Mn, all have some polymerization activity for ethylene or propylene.

The monometallic hypothesis was put on a sound theoretical basis by the quantum chemical theory developed by Cossee,[43] which will be discussed in the next section.

9.5 COSSEE'S MONOMETALLIC MECHANISM

The formation of the active centre and the process of chain growth as described by Cossee were shown in Eq. 3.22 and 3.25 in connection with the kinetic aspects of the mechanism. The mechanism will now be considered in terms of the theoretical treatment of Cossee himself.

In 1960 Cossee,[44] recognizing the important role of transition metal *d*-electrons in many types of heterogeneous catalysis, assumed that the Ti–C bond was the active centre in Ziegler-Natta polymerization. It is well known that many transition metals such as Rh, Pt, Rd and Ni are active in the hydrogenation of olefins, and a number of attempts have been made to correlate the "*d*%-character"† with the catalytic activity. The mechanism developed by Cossee was based on the existence of chlorine vacancies on the $TiCl_3$ surface, as proposed by Arlman.[31] Arlman indicated that such vacancies must exist on the $TiCl_3$ crystal edges in order to ensure the electroneutrality of the crystal, and showed that the number of vacancies is of the order of $10^{-3}/d$ equivalents per mole, where d(expressed in μ) is the crystal dimension perpendicular to the trigonal c axis. For particles having a size well under 1μ, the concentration of vacancies correlates well with the observed number of active sites, as determined by many authors. For

† According to Pauling, "*d* %-character" represents the fraction of *d*-electron participation in the metallic bond. Beeck (*Rev. Mod. Phys.*, **17**, 61, (1945)) was able to show a correlation between the *d* % character of transition metals and their heats of adsorption of ethylene, or their activity in ethylene hydrogenation.

particles of size 1μ, the calculated number of vacancies is 1.4×10^{-3} equiv./mole.

Cossee supposed that active centre formation was due to the interaction of a chlorine vacancy with alkylaluminium, as in Eq. 9.12

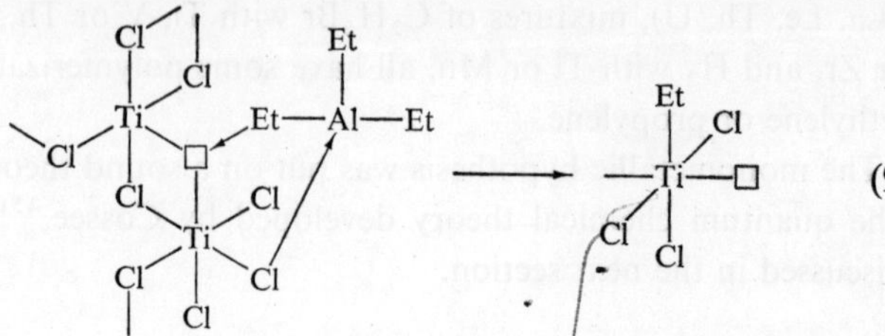

(9.12)

Propagation is due to the adsorption of a monomer unit, which forms a π-bond with Ti, and is then inserted into the Ti–alkyl bond.

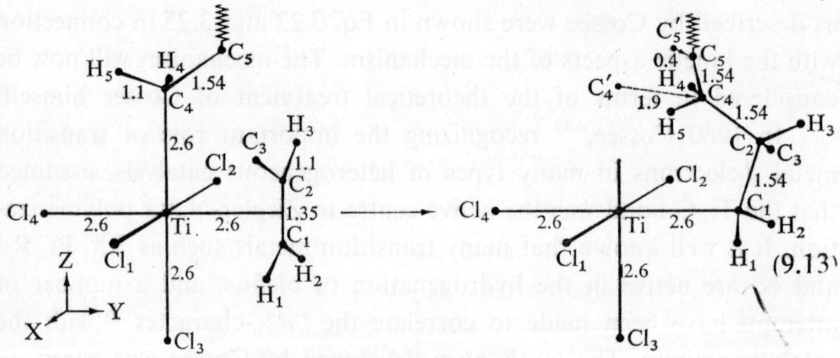

(9.13)

Interatomic distances are shown in angströms. The type of insertion shown in Eq. 9.13 is acceptable for transition metal complexes due to the presence of *d*-orbitals. The d_{yz}-orbital does not affect the Ti-alkyl σ-bond in its equilibrium position, but has a broadening effect on the potential energy curve for angular vibration of the alkyl group. If we consider a transition state in which this alkyl group moves a little over half way along its total reaction path, the orbital of this migrating group overlaps with the d_{z^2}, d_{yz} and p_z orbitals, and the p-orbital of the other olefin already appreciably overlaps the $d_{x^2-y^2}$ orbital, as illustrated in Fig. 9.6. As a result of the availability of *d*-orbitals, the migrating alkyl group has available a combination of orbitals all along its reaction path which allows it to remain relatively strongly bonded to the

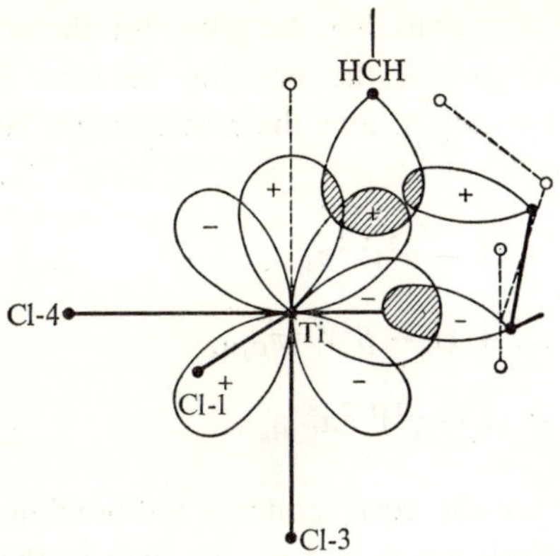

Fig. 9.6. Overlap of the relevant orbitals in the activated polymerization complex according to Cossee.[43]

complex. This ensures that the reaction will have a low energy of activation.

The qualitative exposition given above has been confirmed by the quantum chemical calculations of Cossee[43-45] and Begley and Pannella.[46] According to Cossee,[43] the relevant orbitals in the complex between propylene and the active site, $TiCl_4R$, are given in Fig. 9.7,

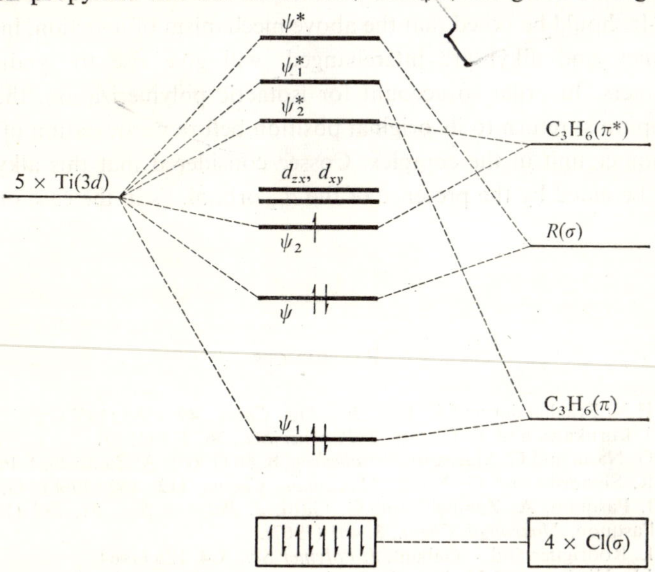

Fig. 9.7. Molecular orbital diagram of the $R\text{-}TiCl_4/C_3H_6$ activated complex.[3]

and the following simplified equations may be given for the bonding orbital between Ti and alkyl group, the bonding between Ti and propylene causing complex formation, and the π-interaction between Ti and propylene, respectively.

$$\psi = \alpha d_{z^2} + (1 - \alpha^2)^{1/2}\sigma_R \tag{9.14}$$

$$\psi_1 = \beta d_{x^2-y^2} + (1 - \beta^2)^{1/2}\pi_{C_3H_6} \tag{9.15}$$

$$\psi_2 = \gamma d_{yz} + (1 - \gamma^2)^{1/2}\pi^*_{C_3H_6} \tag{9.16}$$

The orbitals ψ^*, ψ_1^* and ψ_2^* are the corresponding antibonding combinations. The important orbitals are ψ and ψ_2. As long as the alkyl group, R, is in its equilibrium position, mixing of ψ and ψ_2 cannot occur, but in any non-equilibrium position of R, ψ will be pushed upwards and mixing with ψ_2 starts, so that the upward trend is partially compenstated. This effect is responsible for lowering the energy of insertion. From a detailed consideration of these effects, it has been concluded that the best catalytic activity can be expected when the metal 3*d* orbital (d_{yz}) lies somewhere between π^* and σ_R. It has been shown by Cossee[43)] that this is probably the case in titanium complexes.

It should be noted that the above mechanism of insertion, in which vacancy and alkyl are interchanged, will give rise to syndiotactic polymers. In order to account for isotactic polymerization, the alkyl group must return to its original position before participation of a new monomer unit in the complex. Cossee considered that this alkyl shift may be aided by the presence of the d_{yz} orbital, as in the case of insertion.

References

1. H.N. Friedlander and K. Oita, *Ind. Eng. Chem.*, **49**, 1885 (1957).
2. J. Furukawa and T. Tsuruta, *J. Polymer Sci.*, **36**, 275 (1959).
3. G. Natta and G. Mazzanti, *Tetrahedron*, **8**, 86 (1960); A. Zambelli, I. Pasquon, R. Signorini and G. Natta, *Makromol. Chem.*, **112**, 160 (1968); G. Natta, I. Pasquon, A. Zambelli and G. Gatti, *J. Polymer Sci.*, **51**, 387 (1961); I. Pasquon, *Makromol. Chem.*, **3**, 465 (1967).
4. L. Rodriguez and J. Gabant, *J. Polymer Sci.*, **C4**, 125 (1964).
5. J.P. Hirth and G.M. Pound, *Condensation and Evaporation* (*Progress in Material Science*, **11**), p. 107, Pergammon, 1963.
6. I. Okura and T. Keii, unpublished paper.

7. C.W. Hock, *J. Polymer Sci.*, part A1, **4**, 3055 (1966).
8. I. Okura, Y. Sendoda and T. Keii, *Kogyo Kagaku Zasshi* (Japanese), **73**, 276 (1970); I. Okura, T. Matsumoto, S. Kanetaka, T. Kawai, H. Maeda and T. Keii, presented at Meeting of Japan Cellulose Soc., 1968.
9. J. Boor Jr., *J. Polymer Sci.*, **C1**, 257 (1963).
10. E.J. Arlman, *J. Catalysis*, **3**, 89 (1964).
11. C.D. Nenitzescu and C.A. Hugh, *Angew. Chem.*, **68**, 438 (1956).
12. D.B. Ludlum, A.W. Anderson and C.E. Ashby, *J. Amer. Chem. Soc.*, **80**, 1380 (1958).
13. J. C. McGowan and B.M. Ford, *J. Chem. Soc.*, **1958**, 1149.
14. A.V. Topchev, B.A. Krentsel and L.G. Sidovava, *Dokl. Akad. Nauk SSSR*, **128**, 732 (1959).
15. A.V. Topchev, B.A. Krentsel and L.L. Stotskaya, *Russ. Chem. Rev.*, **30**, 192 (1961).
16. J. Lal. *J. Polymer Sci.*, **31**, 179 (1959).
17. G. Natta, G. Dellasta, G. Mazzanti, U. Glanniri and S. Cesca, *Angew. Chem.*, **71**, 205 (1959).
18. G. Natta and I. Pasquon, *Advan. Catalysis*, **11**, 1 (1959).
19. K. Soga and T. Keii, *J. Polymer Sci.*, part A1, **4**, 2429 (1966).
20. A.P. Firsov and N.M. Chirkov, *Izv. Akad. Nauk SSSR, Ser. Khim.*, **11**, 741 (1964).
21. G. Natta, P. Corradini and I.W. Bassi, *J. Amer. Chem. Soc.*, **80**, 755 (1965).
22. G. Natta, P. Pino, G. Mazzanti, U. Giannini, M. Mantica and M. Peraldo, *J. Polymer Sci.*, **26**, 120 (1957).
23. F. Patat and H. Sinn, *Chim. Ind.* (*Milan*), **44**, 474 (1962).
24. H. Uelzmann, *J. Polymer Sci.*, **32**, 457 (1958).
25. M.L. Huggins, *J. Polymer Sci.*, **48**, 473 (1960).
26. J. Boor Jr., *J. Polymer Sci.*, **C1**, 257 (1963).
27. G.L. Karapinka and W.L. Carrick, *J. Polymer Sci.*, **55**, 145 (1961).
28. G. Henrici-Olive and S. Olive, *Angew. Chem.*, **79**, 764 (1967).
29. G. Henrici-Olive and S. Olive, *J. Organometal. Chem.*, **16**, 339 (1969).
30. C. Beerman and H. Bestian, *Angew. Chem.*, **71**, 618 (1959).
31. J. Arlman, *Proc. Intern. Congr. Catalysis* 3*rd*., **2**, 957 (1964).
32. A. Yamamoto, K. Morifuji, S. Ikeda, T. Saito, Y. Uchida and A. Misono, *J. Amer. Chem. Soc.*, **87**, 4652 (1965); *ibid.*, **90**, 1878 (1968); T. Saito, Y. Uchida, A. Misono, A. Yamamoto, K. Morifuji and S. Ikeda, *J. Org. Chem.*, **6**, 572 (1966); *ibid.*, **88**, 5198 (1966).
33. G. Wilke and J. Herrman, *Angew. Chem.*, **78**, 591 (1966).
34. W.P. Long and D.S. Breslow, *J. Amer. Chem. Soc.*, **82**, 1953 (1960).
35. A.E. Shilov, A.K. Shilova and B.N. Vobkov, *Vysokomolekul. Soedin.*, **4**, 1688 (1962).
36. K.H. Reichert and E. Schubert, *Makromol. Chem.*, **123**, 58 (1969).
37. F.J. Karol and W.L. Carrick, *J. Amer. Chem. Soc.* **83**, 2654 (1961).
38. F. Danusso and D. Sianesi, *Chim. Ind.* (*Milan*), **44**, 474 (1962).
39. A. Schindler, *J. Polymer Sci.*, **B3**, 147 (1965).
40. T. Keii and A. Kojima, unpublished work; A. Kojima, *M. Eng. Thesis, Tokyo Inst. Technology*, 1968.
41. J. Boor Jr., *Polymer Letters*, **2**, 265 (1964); E.A. Youngman and J. Boor Jr., *ibid.*, **4**, 913 (1966); *Amer. Chem. Soc. Div. Polymer Chem.*, **6**, 890 (1965).
42. A.S. Matlack and D.S. Breslow, *J. Polymer Sci.*, part A4, **3**, 2853 (1965).
43. P. Cossee, *J. Catalysis*, **3**, 80, 90 (1964).
44. P. Cossee, *Tetrahedron Letters*, **17**, 12 (1960).
45. P. Cossee and J.H. Schactschneider, presented at 14th Int. Congr. Catalysis, 1968.
46. J.W. Begley and E. Penella, *J. Catalysis*, **8**, 203 (1967).

Acknowledgements

Permission to reproduce the following copyright material is gratefully acknowledged.

Academic Press Inc., New York.
"Advances in Catalysis" vol. 11, p. 1 (1959), (G. Natta and I. Pasquon) Fig. 11 (*Fig. 2.36*) Fig. 12 (*Fig. 2.4*) Fig. 14 (*Fig. 1.2*) Fig. 16 (*Fig. 2.17*) Fig. 17 (*Fig. 2.2*) Fig. 18 (*Fig. 2.18*) Fig. 19 (*Fig. 2.3*) Fig. 20 (*Fig. 2.19*) Fig. 37 (*Fig. 4.9*) Fig. 38 (*Fig. 4.10*)

vol. 19, p. 226 (1967), (M.N. Berger, G.Boocock and R.N. Haward) Fig. 10b (*Fig. 5.26*) Fig. 11 (*Fig. 5.29*) Fig. 12b (*Fig. 1.9*) Fig. 13 (*Fig. 1.10*)

Journal of Catalysis, vol. 2, p. 237 (1963), (A. Clark and G. Bailey) Fig. 2 (*Fig. 6.11*)

The American Chemical Society, New York. "Industrial and Engineering Chemistry." vol. 49, p. 1890 (1957), (H. N. Friedlander) Fig. 10 and 11 (*Fig. 8.11*)

Gauthier-Villars editeur, Paris. "Les Comptes Rendus de L' Academie des Sciences" tome 258, p. 4272 (1964), (G. Lefebre et al.) Fig.(Fig. *2.13*)

Hüthig & Wepf Verlag, Basel. "Die Makromolekulare Chemie" vol. 83, pp. 84, 85, 90, 93, 95 (1965), (M. N. Berger and B. M. Grieveson) Fig. 2 (*Fig. 1.7*) Fig. 3 (*Fig. 1.8*) Fig. 9 (*Fig. 5.17*) Fig. 14 (*Fig. 5.18*) Fig. 16 (*Fig. 5.19*)

vol. 84, p. 99, 101 (1966), (B. M. Grievson) Fig. 3 (*Fig. 5.27*) Fig. 4 (*Fig. 5.28*)

vol. 93, p. 148 (1966), (A. Schindler) Fig. 1 (*Fig. 5.32*)

vol. 70, p. 101 (1964), (A. Schindler) Fig. 2 (*Fig. 5.30*)

John Wiley & Sons, Inc., New York. "Journal of Polymer Science" Al, 4, p. 2781 (1966), (A. K. Ingberman *et al.*) Fig. 2 (*Fig. 2.24*) Fig. 3 (*Fig. 2.25*) Fig. 4 (*Fig. 4.17*) Fig. 8 (*Fig. 1.4*)

C4, p. 215 (1964), (W. Cooper *et al.*) Fig. 3 (*Fig. 1.1*)

C4, p. 72 (1964), (H. Schnecko *et al.*) Fig. 1 (*Fig. 2.22*) Fig. 2 (*Fig. 2.23*)

C4, p. 125 (1964), (L. Rodriguez *et al.*) Fig. (*Fig. 9.4*)

C4, p. 84 (1964), (A. Schindler) Fig. 2 (*Fig. 5.33*)

C16, p. 425 (1967), (J. Ambroz *et al.*) Fig. 1 (*Fig. 3.4*) Fig. 3 (*Fig. 3.3*)

"Polymerization by Organometallic Compounds" by L. Reich and A. Schindler, p. 320, Fig. IV–37 (Fig. 5.31).

The Koninklijke Nederlandse Chemische Vereniging, Hague.
Recueil des travaux Chimiques des Pays-Bas, vol. 81, p. 233 (1962), (H. J. M. Bartelink *et al.*) Fig. 5 (*Fig. 8.10*)

Marcel Dekker, Inc., New York. "The Stereochemistry of Macromolecules" ed. by A. D. Ketley, vol. 1, p. 160, (1967) (P. Cossee) Fig. 6 and 7 (*Fig. 9.5*)

Organo Ufficiale della Societa Chimica-Italiana, Milan. "La Chimica e l'Industria" vol. 44, p. 621 (1962), (G. Natta *et al.*) Fig. 1, (*Fig. 2.20*)

Figure numbers in italics give locations in this book.

Author Index

Numbers in parentheses are reference numbers and indicate that an author's work is referred to although his name may not be cited in the text. Numbers in bold show the page on which the complete reference is listed.

K

L

M

T

U

V

Subject Index

N

O

P

Q

R

S

T

U

V

Z